Lecture Notes in Mathematics

A collection of informal reports and seminars
Edited by A. Dold, Heidelberg and B. Eckmann, Zürich

T0233349

272

K. R. Parthasarathy
University of Bombay, Bombay/India
K. Schmidt
Bedford College, University of London, London/England

Positive Definite Kernels, Continuous Tensor Products, and Central Limit Theorems of Probability Theory

Springer-Verlag
Berlin · Heidelberg · New York 1972

AMS Subject Classifications (1970): 43A05, 43A35, 43A65, 46C10, 60B15, 60F05

ISBN 3-540-05908-3 Springer-Verlag Berlin · Heidelberg · New York
ISBN 0-387-05908-3 Springer-Verlag New York · Heidelberg · Berlin

This work is subject to copyright. All rights are reserved, whether the whole or part of the material is concerned, specifically those of translation, reprinting, re-use of illustrations, broadcasting, reproduction by photocopying machine or similar means, and storage in data banks.

Under § 54 of the German Copyright Law where copies are made for other than private use, a fee is payable to the publisher, the amount of the fee to be determined by agreement with the publisher.

© by Springer-Verlag Berlin · Heidelberg 1972. Library of Congress Catalog Card-Number 72-85400. Printed in Germany.

Offsetdruck: Julius Beltz, Hemsbach/Bergstr.

PREFACE

These notes are mainly based on a course of lectures given by the first named author at the Research and Training School of the Indian Statistical Institute, Calcutta during May 1971. A first and slightly shorter version of these notes has appeared under the same title as publication No. M71-1 of the Research and Training School of the Indian Statistical Institute. Some of the results were obtained when the authors were at the Statistical Laboratory, Mathematics Department, University of Manchester, in 1970.

The notion of a continuous tensor product of Hilbert spaces and group representations appears in the work of H. Araki [2], H. Araki and E.J. Woods [1] and R.F. Streater [19]. Their analysis leads to a connection between continuous tensor products and the theory of infinitely divisible distributions of Probability theory. The present work contains a systematic study of these notions in terms of positive definite kernels with invariance properties under a group action. Such analysis also leads to a unified approach to the central limit problems of Probability theory, the theory of stochastic processes with stationary increments and construction of free fields in Quantum Mechanics.

The contents of these notes are divided into three parts. In part 1 the notion of a projectively invariant positive definite kernel on an abstract G-space is introduced and obtained as expectation value of a projective unitary representation of the group G. Affine invariant conditionally positive definite kernels are investigated and a representation of such kernels in terms of unitary representations and first order cocycles is obtained. Using these ideas and the theory of multiplicative measures, continuous tensor products of Hilbert spaces and representations are constructed. The Fock-Cook construction of the Bose-Einstein field in Quantum Mechanics [4] , [18] arises as a natural consequence of this theory. Much of the inspiration for this approach was derived from the work of H. Araki [2] .

Part 2 analysis limits of products of uniformly infinitesimal families of positive definite kernels. This leads in particular to the limit laws in the theory of sums of independent random variables.

The third part contains results about first order cocycles for various classes of representations and on some special groups. In particular it is shown that first order cocycles of induced representations arise from the cocycles of the inducing representation. As corollary to this, cocycles of irreducible representations of nilpotent Lie groups are described. Cocycles of representations of semisimple Lie groups which are induced by characters of maximal solvable subgroups are also obtained.

A short list of references including most of the major contributions to these problems is included.

The first named author would like to thank the Research and Training School of the Indian Statistical Institute and the University of Bombay for providing him facilities for writing these notes. For the same reason the second author would like to express his gratitude to the Department of Mathematics, Bedford College, University of London.

K.R. Parthasarathy
K. Schmidt

CONTENTS

Part I: Positive definite kernels and continuous tensor products

§1. Positive definite and conditionally positive definite kernels

Here we shall concern ourselves with some elementary properties of positive definite kernels.

Definition 1.1. Let X be any set. A complex valued function $K(.,.)$ on $X\times X$ is called a hermitian kernel if $K(x,y) = \overline{K(y,x)}$ for all $x,y\in X$. A hermitian kernel is said to be positive definite if, for any positive integer n and for every choice of elements $x_1,\dots,x_n$ in X and complex numbers $a_1,\dots,a_n$, the inequality

$$\sum_{j,k=1}^{n} a_j\overline{a_k}K(x_j,x_k) \geq 0 \qquad (1.1)$$

holds. K is said to be conditionally positive definite if the above inequality holds whenever $\sum_j a_j = 0$.

Theorem 1.2. Let X be a set and $K(.,.)$ be a positive definite kernel on $X\times X$. Then there exists a complex Hilbert space H and a map v from X into H such that the set of vectors $\{v(x), x\in X\}$ spans H and

$$K(x,y) = \langle v(x),v(y)\rangle \text{ for all } x,y\in X,$$

where $\langle .,. \rangle$ stand for the inner product in H.

Proof: If $x_1,\dots,x_n$ are any n points in X, then the matrix $((K(x_j,x_k)))$ is positive definite. Hence there exists an n dimensional complex Gaussian probability distribution $\mu^{x_1\dots x_n}$ with mean vector zero and covariance matrix $((K(x_j,x_k)))$.

The family of probability measures $\mu^{x_1 \ldots x_n}$ is obviously consistent. Consider the borel space $\mathfrak{X}$ of all complex valued functions on X with the smallest σ-field relative to which every projection map Π_x: $f \to f(x)$ from $\mathfrak{X}$ into the complex plane is measurable. By Kolmogorov's theorem there exists a measure μ on $\mathfrak{X}$ such that the joint distribution of $(f(x_1),\ldots,f(x_n))$, $f \in \mathfrak{X}$, is $\mu^{x_1,\ldots,x_n}$ for every $x_1,\ldots,x_n \in X$. If we consider the Hilbert space $L_2(\mu)$ and define $v(x) = f(x)$, $f \in \mathfrak{X}$, then $v(x)$ is an element of $L_2(\mu)$ and

$$\langle v(x),v(y)\rangle = \int f(x)\overline{f(y)}\, d\mu(f) = K(x,y).$$

If we define H as the subspace spanned by the vectors $v(x), x \in X$, the proof of the theorem is complete.

<u>Corollary 1.3</u>. If X is a topological space and $K(.,.)$ is a continuous positive definite kernel on X×X, then the map v of Theorem 1.2 is continuous. In fact

$$||v(x) - v(y)|| = \{K(x,x) + K(y,y) - 2 \operatorname{Re} K(x,y)\}^{\frac{1}{2}}.$$

<u>Corollary 1.4</u>. If X is a topological space and $K(.,.)$ is a positive definite kernel on X×X, then K is continuous if and only if its real part is continuous. In fact, $K(x,x)$ is real and we have the following inequality:

$$|K(x,y) - K(x',y')| \leq K(x,x)^{\frac{1}{2}}.\{K(y,y) + K(y',y') - 2\,\mathrm{Re}\,K(y,y')\}^{\frac{1}{2}}$$
$$+ K(y',y')^{\frac{1}{2}}.\{K(x,x) + K(x',x') - 2\,\mathrm{Re}\,K(x,x')\}^{\frac{1}{2}}. \tag{1.2}$$

Lemma 1.5. If $A = ((a_{jk}))$, $B = ((b_{jk}))$ are two positive definite n-th order matrices, then the matrix $C = ((a_{jk}b_{jk}))$ is also positive definite.

Proof: Let $X = (X_1,\ldots,X_n)$ and $Y = (Y_1,\ldots,Y_n)$ be two independent complex normally distributed random variables with mean vector zero and covariance matrices A and B respectively. Then C is the covariance matrix of the random vector $Z = (X_1Y_1,\ldots, X_nY_n)$. This shows that C is positive definite and completes the proof.

Lemma 1.6. Let X be a set and K_1 and K_2 be two positive definite kernels on X×X. Then K_1K_2 is also positive definite. If a and b are non-negative constants then aK_1+bK_2 is also positive definite.

Proof: That K_1K_2 is positive definite follows from Lemma 1.5. The second part is trivial.

Lemma 1.7. Let X be a set and L a hermitian kernel on X×X. The following conditions are equivalent:

(1) L is conditionally positive definite,

(2) for any fixed $x_0 \varepsilon X$, the kernel L_{x_0} defined by

$$L_{x_0}(x,y) = L(x,y)-L(x,x_0)-L(x_0,y)+L(x_0,x_0)$$

for all $x,y \varepsilon X$, is positive definite,

(3) for every $t>0$, $\exp tL$ is positive definite.

<u>Proof</u>: Let L be conditionally positive definite. Let $x_1,\dots,x_n$ be any n points in X and $a_1,\dots,a_n$ be complex numbers. Putting $a_o = -\sum_{j=1}^{n} a_j$, we have from Definition 1.1,

$$\sum_{j,k=0}^{n} a_j\overline{a_k}\ L(x_j,x_k) \geq 0.$$

Rewriting the above inequality, we have

$$\sum_{j,k=1}^{n} a_j\overline{a_k}\ \{L(x_j,x_k)-L(x_j,x_o)-L(x_o,x_k)+L(x_o,x_o)\} \geq 0.$$

This shows that (1) implies (2).

Suppose now (2) holds. Then by Lemma 1.6, exp tL_{x_o} is a positive definite kernel for every $t>0$. Since any kernel of the form $K(x,y) = k(x)\overline{k(y)}$ is positive definite, it follows once again from Lemma 1.6 that

$$\exp tL(x,y) = \exp -tL(x_o,x_o).\exp t(L(x,x_o)+L(x_o,y)).\exp tL_{x_o}(x,y)$$

is positive definite Thus (2) implies (3).

If (3) holds, then we observe that the kernel $t^{-1}(\exp tL - 1)$ is conditionally positive definite for every $t>0$. Letting $t\to 0$, we see that the limit L is also conditionally positive definite. This shows that (3) implies (1) and completes the proof of the lemma.

<u>Remark 1.8</u>. If X is the real line and L is a conditionally positive definite kernel of the form $L(x,y) = \psi(x-y)$, where ψ

is a continuous function, then exp $t\psi$ is an 'infinitely divisible positive definite function' in the sense of Levy and Khinchine. They are Fourier transforms of infinitely divisible probability distributions on the real line. The reader may refer to B.V. Gnedenko and A.N.Kolmogorov, Limit distributions of sums of independent random variables, chapter 3.

§2. Projectively invariant positive definite kernels

Here we shall study some properties of positive definite kernels which satisfy certain invariance conditions under a group of transformations.

Let G be a group with identity element e and let X be a set. We say the group G acts on X if there exists a map $(g,x) \to gx$ from $G \times X$ onto X such that the following conditions are satisfied:

(1) for every fixed g, the map $x \to gx$ is one to one and onto.

(2) $ex = x$ for all $x \in X$,

(3) $g_1(g_2x) = g_1g_2x$ for all $g_1, g_2 \in G$, $x \in X$.

G acts transitively on X, if for any $x, y \in X$ there exists a $g \in G$ such that $gx = y$. If G is a topological group and X is a topological space we say that G acts continuously on X if the map $(g,x) \to gx$ is continuous and for each fixed g, the map $x \to gx$ is a homeomorphism of X.

Definition 2.1. Let G be a group acting on X. A hermitian kernel K on $X \times X$ is said to be projectively invariant under G if

$$K(gx,gy) = \alpha(g,x)\overline{\alpha(g,y)}K(x,y) \text{ for all } x,y\in X, g\in G, \quad (2.1)$$

for some complexvalued function $\alpha(g,x)$ defined on $G\times X$. K is said to be invariant under G if $\alpha(g,x) \equiv 1$ in (2.1). The function $\alpha(.,.)$ is called the multiplier of K. Without loss of generality we may assume that $\alpha(e,x) = 1$ for all $x\in X$.

Lemma 2.2. Let $\alpha(.,.)$ be the multiplier of a projectively invariant positive definite kernel K. Then there exists a function $\sigma(.,.)$ on $G\times G$ satisfying the following properties:

(1) $\alpha(g_1g_2,x) = \sigma(g_1,g_2)\alpha(g_1,g_2x)\alpha(g_2,x)$,

(2) $|\sigma(g_1,g_2)| = \sigma(g_1,e) = \sigma(e,g_2) = 1$,

(3) $\sigma(g_1,g_2)\sigma(g_1g_2,g_3) = \sigma(g_1,g_2g_3)\sigma(g_2,g_3)$

for all $g_1,g_2,g_3\in G$.

Proof: Let $X_o = \{x:K(x,x)\neq 0\}$. Since (2.1) implies that

$$K(x,x) = |\alpha(g^{-1},gx)|^2 . K(gx,gx),$$

$$K(gx,gx) = |\alpha(g,x)|^2 . K(x,x),$$

it follows that $gx\in X_o$ whenever $x\in X_o$. Further $\alpha(g,x)\neq 0$ for all $g\in G$, $x\in X_o$. From (2.1) we have

$$\alpha(g_1g_2,x)\overline{\alpha(g_1g_2,y)} = \alpha(g_1,g_2x)\alpha(g_2,x)\overline{\alpha(g_1,g_2y)\alpha(g_2,y)} \quad (2.2)$$

for all $x,y\in X_o, g_1,g_2\in G$.

Gathering all terms containing x on one side and those containing y on the other side in equation (2.2), we see that there exists a constant $\sigma(g_1,g_2)$ such that property (1) of the lemma is satisfied for all $x\varepsilon X_o$. If $x\notin X_o$, both sides of (1) vanish. Putting x=y in (2.2) we obtain $|\sigma(g_1,g_2)| = 1$. Since $\alpha(e,x)\equiv 1$, it follows that $\sigma(g_1,e) = \sigma(e,g_2) = 1$. Writing $\alpha(g_1g_2g_3,x)$ in two different ways by considering $g_1g_2g_3$ as $g_1(g_2g_3)$ and $(g_1g_2)g_3$ we obtain property (3). This completes the proof of the lemma.

<u>Definition 2.3</u>. Let G and H be groups with identity elements e and e' respectively. A map σ from G×G into H is called a <u>second order cocycle</u> with values in H if the following identities are satisfied:

$$\sigma(g,e) = \sigma(e,g) = e' \text{ for all } g\varepsilon G,$$

$$\sigma(g_1,g_2)\sigma(g_1g_2,g_3) = \sigma(g_1,g_2g_3)\sigma(g_2,g_3) \quad \text{for all } g_1,g_2,g_3\varepsilon G.$$

<u>Remark 2.4</u>. The function σ appearing in Lemma 2.2 is a second order cocycle with values in the multiplicative group T of complex numbers of modulus unity (the circle group).

<u>Definition 2.5</u>. Let G be a group and H a complex Hilbert space. By a projective representation of G in H we mean a map $g\to U_g$ from G into the group of unitary operators in H, satisfying

$$U_e = I, \text{ where I is the identity operator in H,} \qquad (2.3)$$

$$U_{g_1}U_{g_2} = \sigma(g_1,g_2).U_{g_1g_2} \text{ for all } g_1,g_2\varepsilon G, \text{ where } \sigma \text{ is a}$$

scalar valued function on G×G.

Remark 2.6. Equations (2.3) and the unitarity of U imply that σ is a second order cocycle with values in the torus T. This follows from writing $U_{g_1}U_{g_2}U_{g_3}$ in two different ways.

We are now ready to state the main result of this section:

Theorem 2.7. Let X be a set and G a group acting on it. Suppose K is a positive definite kernel on X×X which is projectively invariant under G. Then there exists a Hilbert space H with inner product $<.,.>$, a projective representation $g \to U_g$ of G in H and a map $v: X \to H$ such that the vectors $\{v(x),\ x \in X\}$ span H and that

$$K(x,y) = <v(x),v(y)> \text{ for all } x,y \in X, \tag{2.4}$$

$$v(gx) = \alpha(g,x)U_g v(x) \text{ for all } g \in G,\ x \in X, \tag{2.5}$$

where α is the multiplier of K. Conversely, any kernel K satisfying (2.4) and (2.5) for some projective representation U is projectively invariant and positive definite.

Proof: By Theorem 1.2 we can construct the Hilbert space H and the map $v: X \to H$ such that (2.4) is satisfied. Let X_o be the set $\{x: K(x,x) \neq 0\}$. Then $v(x) \neq 0$ and $\alpha(g,x) \neq 0$ for all $g \in G$, $x \in X_o$, exactly as in the proof of Lemma 2.2. We define a map $U_g: v(x) \to \alpha(g,x)^{-1}v(gx)$ on the set $\{v(x),\ x \in X_o\}$. Since K satisfies (2.1) and (2.4), we have

$$<U_g v(x), U_g v(y)> = <v(x),v(y)> \text{ for all } x,y \in X_o.$$

Since the vectors $\{v(x),\ x \in X_o\}$ span H, U_g can be linearly extended to an isometry on H. Since the extended isometry is onto, it is actually a unitary mapping. We denote this unitary operator again by U_g. From Lemma 2.2 we have

$$U_{g_1}U_{g_2}\, v(x) = \alpha(g_1g_2,x)\alpha(g_1,g_2x)^{-1}\alpha(g_2,x)^{-1}U_{g_1g_2}v(x)$$

$$= \sigma(g_1,g_2)U_{g_1g_2}v(x) \qquad \text{for all } x\varepsilon X_o.$$

Hence

$$U_{g_1}U_{g_2} = \sigma(g_1,g_2)U_{g_1g_2} \qquad \text{for all } g_1,g_2\varepsilon G.$$

This proves the first part. The converse is trivially verified. The theorem is proved completely.

Remark 2.8. If K is an invariant positive definite kernel, then the projective representation U of the above theorem can be replaced by an ordinary representaion, i.e. σ can be taken as identically equal to 1.

Remark 2.9. If G is a topological group and X is a topological space such that G acts continuously on X and K is a continuous positive definite kernel on X×X, then the map $x\to v(x)$ of the theorem is continuous and the representation $g\to U_g$ is weakly continuous, i.e. for any two vectors $v_1,v_2\varepsilon H$, $\langle U_g v_1,v_2\rangle$ is a continuous function on G.

Remark 2.10. If X=G, G acts on G by left multiplication, and K is an invariant positive definite kernel on G×G, then K is of the form $K(g_1,g_2) = \phi(g_2^{-1}g_1)$ for some complex valued function ϕ on G and all $g_1,g_2\varepsilon G$. If we write $v_o=v(e)$, it follows that

$\phi(g) = \langle U_g v_o, v_o \rangle$ for all $g \in G$. This is a classical result of Gelfand and Raikov. We define therefore:

Definition 2.11. Let G be a group and ϕ a complex valued function on G. ϕ is called positive definite, if the kernel K defined by $K(g,h) = \phi(h^{-1}g)$ for all $g,h \in G$, is positive definite. ϕ is said to be normalized, if $\phi(e) = 1$.

§3. Affine invariant conditionally positive definite kernels

Definition 3.1. Let G be a group acting on a set X. A hermitian kernel L on X×X is said to be affine invariant under G if

$$L(gx,gy) = \beta(g,x) + \overline{\beta(g,y)} + L(x,y) \quad \text{for all } x,y \in X,\ g \in G, \tag{3.1}$$

for some complexvalued function $\beta(g,x)$ defined on G×X. L is said to be invariant under G if $\beta \equiv 0$ in (3.1).

Lemma 3.2. Let L be an affine invariant conditionally positive definite kernel satisfying (3.1). Then there exists a second order cocycle s on G×G with values in the additive group of real numbers, satisfying the following equation:

$$\beta(g_1g_2,x) - \beta(g_1,g_2x) - \beta(g_2,x) = is(g_1,g_2) \quad \text{for all } g_1,g_2 \in G,\ x \in X.$$

Proof: This lemma is proved in the same manner as Lemma 2.2.

Definition 3.3. Let G be a group acting on X and let $g \to U_g$ be a unitary representation of G in a Hilbert space H. A map $v: X \to H$ is called a first order cocycle or simply a cocycle with values in H and origin $x_o \in X$ if

$$U_g v(x) = v(gx) - v(gx_o) \quad \text{for all } g \in G,\ x \in X. \tag{3.2}$$

Theorem 3.4. Let G be a group acting on X and let L be an affine invariant conditionally positive definite kernel on X×X. Then for any fixed $x_o \in X$, there exists a unitary representation $g \to U_g$ of G in a Hilbert space H and a cocycle v with values in H and origin x_o such that

$$\langle v(x), v(y)\rangle = L(x,y) - L(x,x_o) - L(x_o,y) + L(x_o,x_o) \tag{3.3}$$
$$\text{for all } x,y \in X.$$

Conversely, if U is a unitary representation of G in H and v is a cocycle with values in H and origin x_o, and if L is a kernel satisfying (3.3), then L is an affine invariant conditionally positive definite kernel.

Proof: Let L be an affine invariant conditionally positive definite kernel. Then by Lemma 1.7, the left hand side of (3.3) is positive definite. Hence by Theorem 1.2, there exists a Hilbert space H and a map $v: X \to H$ such that (3.3) is satisfied. Let $A = \{v(x) - v(y),\ x,y \in X\}$. On the subset A of H we define a map U_g by the equation

$$U_g\{v(x) - v(y)\} = v(gx) - v(gy), \text{ for all } x,y \in X. \tag{3.4}$$

Since L is affine invariant, (3.1), (3.3) and (3.4) imply that

$$\langle U_g\{v(x) - v(y)\}, U_g\{v(x') - v(y')\}\rangle =$$

$$\langle v(x) - v(y), v(x') - v(y')\rangle \quad \text{for all } x,x',y,y' \in X.$$

Thus U_g is an isometry on the set A. Hence U_g can be extended linearly to an isometry on the closed linear span of A. Since (3.3) implies that $v(x_o) = 0$, A spans H, and the image of H under the extended U_g is also H. Hence U_g is a unitary operator on H. (3.4) implies that $U_{g_1} U_{g_2} = U_{g_1 g_2}$ for all $g_1, g_2 \in G$. Putting $y = x_o$ in (3.4) we obtain the equation (3.2). Thus v is a cocycle with origin x_o. This proves the first part. Conversely, if (3.3) is fulfilled, then the left hand side of (3.3) is a positive definite kernel and hence L is conditionally positive definite. The affine invariance of L is an easy consequence of equation (3.2). The theorem is proved completely.

Remark 3.5. If G is a topological group acting continuously on a topological space X and if L is a continuous affine invariant conditionally positive definite kernel, then the cocycle v satisfying (3.3) is continuous and the representation U is weakly continuous.

As before we note that any invariant conditionally positive definite kernel L on GxG can be written as $L(g,h) = \psi(h^{-1}g)$ for some complexvalued function ψ on G and all $g,h \in G$. This leads to

Definition 3.6. A complex valued function ψ on a group G is called conditionally positive definite, if the kernel L defined by $L(g,h) = \psi(h^{-1}g)$ for all $g,h \in G$, is conditionally positive definite. ψ is called normalized, if $\psi(e) = 0$.

§4. Multiplicative measures

The main purpose of this section is to establish that multiplicative measures are, under certain general conditions, exponentials of additive measures.

Definition 4.1. A borel space $(\Omega,\mathfrak{S})$ is said to be standard if it is borel isomorphic to the unit interval with its usual borel structure.

Definition 4.2. Let $(\Omega,\mathfrak{S})$ be a standard borel space. A function $M:\mathfrak{S}\to C_o$, where C_o stands for the space of nonzero complex numbers, is called a non-atomic complex multiplicative measure if the following holds:

(1) $M(\bigcup_{n=1}^{\infty} A_n) = \prod_{n=1}^{\infty} M(A_n)$ for any sequence (A_k) of disjoint borel sets,

(2) $M(\emptyset) = M(\{\omega\}) = 1$ for any single point set $\{\omega\}$, $\omega\in\Omega$.

Let for the following $(\Omega,\mathfrak{S})$ be a fixed standard borel space and M a fixed nonatomic complex valued multiplicative measure on $(\Omega,\mathfrak{S})$.

Lemma 4.3. $\sup_{A\in\mathcal{S}} |M(A)| = \beta(M) < \infty.$ (4.1)

Proof: The map $A\to\log |M(A)|$ defines a signed measure which takes finite value for every borel set. An application of the Hahn-Jordan decomposition theorem gives the result.

Definition 4.4. For any nonatomic complex multiplicative measure M on $(\Omega,\mathcal{S})$, its oscillation $\alpha(A)$ over any $A\in\mathcal{S}$ is defined by

$$\alpha(A) = \sup_{B\subset A,\ B\in\mathcal{S}} |M(B) - 1|.$$

Lemma 4.5. For any sequence (A_n) of disjoint borel sets,

$$\alpha(\bigcup_{n=1}^{\infty} A_n) \leq \beta(M). \sum_{n=1}^{\infty} \alpha(A_n), \tag{4.2}$$

where $\beta(M)$ is defined by (4.1).

Proof: Let (A_n) be a sequence of disjoint borel sets and ε a positive number. We choose a set $D \subset \bigcup_{n=1}^{\infty} A_n$ such that

$$|M(D) - 1| \geq \alpha(\bigcup_{n=1}^{\infty} A_n) - \varepsilon.$$

But

$$|M(D) - 1| = |\prod_{n=1}^{\infty} M(D\cap A_n) - 1| \leq \sum_{n=1}^{\infty} |M(\bigcup_{k=n+1}^{\infty} D\cap A_k)|.$$

$$.|M(D\cap A_n) - 1| \leq \beta(M). \sum_{n=1}^{\infty} \alpha(A_n).$$

Since ε is arbitrary, the above inequalities complete the proof of the lemma.

Lemma 4.6. Let (A_n) be a sequence of borel sets descending to a single point set $\{\omega\}$. Then $\lim_n \alpha(A_n) = 0$.

Proof: Let $B_n = A_n - \{\omega\}$. For every n, choose $D_n \subset B_n - B_{n+1}$, such that

$$|M(D_n) - 1| \geq \alpha(B_n - B_{n+1}) - 2^{-n}.$$

Then we have

$$\sum_{n=1}^{\infty} \alpha(B_n - B_{n+1}) \leq \sum_{n=1}^{\infty} |M(D_n) - 1| + 1. \tag{4.3}$$

Since the infinite product $\prod_{n=1}^{\infty} M(D_n)$ converges to $M(\bigcup_{n=1}^{\infty} D_n)$ irrespective of the order of the D_n's, it follows that $\sum_{n=1}^{\infty} |M(D_n) - 1| < \infty$. Hence the infinite series on the left side of (4.3) converges. Thus by Lemma 4.5,

$$\lim_n \alpha(A_n) = \lim_n \alpha(B_n) \leq \beta(M).\lim_n \sum_{k=n}^{\infty} \alpha(B_n - B_{n+1}) = 0.$$

This completes the proof of the lemma.

Theorem 4.7. Let $(\Omega, \mathcal{S})$ be a standard borel space and M be a nonatomic complex multiplicative measure on $\mathcal{S}$. Then there exists a unique totally finite complex nonatomic measure m on $\mathcal{S}$, such that

$$M(A) = \exp m(A) \quad \text{for all } A \in \mathcal{S}. \tag{4.4}$$

Proof: Since $(\Omega, \mathcal{S})$ is standard, we may assume without loss of generality that it is the closed unit interval with its usual borel structure. If α is the oscillation of M, then it follows from Lemma 4.6 that for every $\omega \in \Omega$, there exists a neighbourhood

$N(\omega)$ of ω such that $\alpha(N(\omega))<\frac{1}{2}$. Using the compactness of the space Ω we can select a finite number of such neighbourhoods $N_1,\dots,N_n$, which cover Ω. Let

$$B_k = N_k - \bigcup_{j=1}^{k-1} N_j,$$

$$m(A) = \sum_{k=1}^{n} \operatorname{Log} M(A \cap B_k),$$

where Log stands for the principal value of the logarithm. Clearly m is a countably additive measure satisfying (4.4). The total finiteness and the uniqueness of m are obvious. This completes the proof of the theorem.

Lemma 4.8. Let $(\Omega,\mathcal{S})$ be a standard borel space and let M be a nonatomic complex multiplicative measure on $\mathcal{S}$. Then we have

$$\sup \sum_{k=1}^{n} |M(A_k) - 1| = \gamma(M) < \infty, \tag{4.5}$$

where the supremum is taken over all finite borel partitions $(A_1,\dots,A_n)$, $n\geq 1$, of Ω.

Proof: Assume that $\gamma(M) = \infty$. Then there exists a partition $(A_1,\dots,A_n)$ of Ω, such that

$$\sum_{k=1}^{n} |M(A_k) - 1| > 2\beta(M).\mathrm{Var}\ m(\Omega),$$

where β is defined by (4.1), m is the measure defined in (4.4), and Var stands for the variation. Let $(B_1,\dots,B_s)$ be a refinement of $(A_1,\dots,A_n)$ such that $\sup_j \alpha(B_j) < \frac{1}{2}$, where α is the oscillation of M. We conclude from Lemma 4.5, that

$$\sum_{k=1}^{n} |M(A_k) - 1| \leq \beta(M). \sum_{k=1}^{s} |M(B_k) - 1|.$$

Together with the inequality

$$2 \, |\text{Log } (1+x)| \geq |x| \quad \text{for } |x|<\tfrac{1}{2}$$

this implies that

$$\sum_{k=1}^{n} |M(A_k) - 1| \leq 2\beta(M). \sum_{k=1}^{s} |m(B_k)| \leq 2\beta(M).\text{Var } m(\Omega),$$

which contradicts our assumption. This proves the lemma.

Remark 4.9. It is easy to see that $\gamma(M)(A) = \sup \sum_{k=1}^{n} |M(A_k) - 1|$, where the supremum is taken over all finite borel partitions $(A_1,\ldots,A_n)$, $n\geq 1$, of A, is a measure on $\mathcal{S}$, which coincides with Var m.

§5. Factorisable families of projectively invariant positive definite kernels

Definition 5.1.
Let X be a set, $(\Omega,\mathcal{S})$ a standard borel space and $\{f(A,.),A\in\mathcal{S}\}$ a family of complex valued functions on X. The family $\{f(A,.), A\in\mathcal{S}\}$ is said to be factorisable if, for every $x\in X$, $f(.,x)$ is a nonatomic complex multiplicative measure on $\mathcal{S}$. $\{f(A,.),A\in\mathcal{S}\}$ is called an additive family if $f(.,x)$ is a totally finite complex valued measure on $\mathcal{S}$ for every fixed $x\in X$.

Theorem 5.2. Let X be a set, G a group acting on X and $(\Omega,\mathcal{S})$ a standard borel space. Suppose $\{K(A,.,.), A\in\mathcal{S}\}$ is a factorisable family of projectively invariant positive definite kernels on $X \times X$. Then there exists a unique additive family $\{L(A,.,.), A\in\mathcal{S}\}$ of affine invariant conditionally positive definite kernels on $X \times X$ such that, for every $A \in \mathcal{S}$, $x,y \in X$,

$$K(A,x,y) = \exp L(A,x,y), \tag{5.1}$$

and $L(.,x,y)$ is nonatomic for every x,y. If each kernel $K(A,.,.)$ is invariant, then the same is true for each $L(A,.,.)$.

Proof: For every fixed $x,y \in X$ we construct the unique measure $L(.,x,y)$ according to Theorem 4.7 so that (5.1) is fulfilled for every $A \in \mathcal{S}$. To prove the conditional positive definiteness of $L(A,.,.)$ consider any n points $x_1,\ldots,x_n \in X$. Let $\{A_{rs}, 1\leq s\leq r\}$, $r=1,2,\ldots$, be a family of borel partitions of A such that

$$\lim_r \sup_s (\mathrm{Var}\ L)(A_{rs},x_j,x_k) = 0 \text{ for } j,k = 1,\ldots,n,$$

where Var stands for variation. Since

$$|e^x - 1 - x| \leq |x|^2 \quad \text{for } |x|<1,$$

we have

$$\lim_r \left|L(A,x_j,x_k) - \sum_{s=1}^{r} \{K(A_{rs},x_j,x_k) - 1\}\right|$$

$$\leq \lim_r \sum_{s=1}^{r} |L(A_{rs},x_j,x_k) - K(A_{rs},x_j,x_k) + 1|$$

$$\leq \mathrm{Var}\ L(A,x_j,x_k).\lim_r \sup_s |L(A_{rs},x_j,x_k)| = 0 \tag{5.2}$$

The positive definiteness of K implies that for all complex numbers $a_1,\ldots,a_n$ with $\sum_{j=1}^{n} a_j = 0$,

$$\sum_{s=1}^{r} \sum_{j,k=1}^{n} a_j\overline{a_k} \{K(A_{rs},x_j,x_k) - 1\} \geq 0.$$

Hence (5.2) implies that L(A,.,.) is conditionally positive definite. To complete the proof we observe that the projective invariance of K(A,x,y) implies that

$$K(A,gx,gy) = \alpha(A,g,x)\overline{\alpha(A,g,y)}\, K(A,x,y) \tag{5.3}$$

for all $A\in\mathcal{S}$, $g\in G$, $x,y\in X$, where α is a complex valued function. Since K(.,gx,gy) and K(.,x,y) are nonatomic multiplicative measures, it follows that $\alpha(.,g,x)\overline{\alpha(.,g,y)}$ is a nonatomic multiplicative measure for every fixed $g\in G$, $x,y\in X$. If we choose and fix a point $x_o\in X$ and put

$$P(A,g) = \frac{\overline{\alpha(A,g,x_o)}}{|\alpha(A,g,x_o)|}$$

it follows that the function

$$\alpha'(A,g,x) = \alpha(A,g,x).P(A,g)$$

is a nonatomic multiplicative measure in A. Since $|P(A,g)| = 1$, (5.3) can be written as

$$K(A,gx,gy) = \alpha'(A,g,x)\overline{\alpha'(A,g,y)}\, K(A,x,y).$$

The affine invariance follows now from (5.1) and an application of Theorem 4.7 to α'. From our construction it is immediate that if K is invariant, then L is invariant as well. This completes the proof of the theorem.

Finally we shall consider the case of continuous kernels on a topological space.

<u>Theorem 5.3</u>. Let G be a topological group acting continuously on a topological space X, and $(\Omega, \mathcal{S})$ a standard borel space. Suppose $\{K(A,.,.), A\varepsilon\mathcal{S}\}$ is a factorisable family of continuous projectively invariant positive definite kernels on X×X. Then there exists a factorisable family $\{\alpha(A,.), A\varepsilon\mathcal{S}\}$ of continuous functions on X and an additive family $\{M(A,.,.), A\varepsilon\mathcal{S}\}$ of continuous affine invariant positive definite kernels, satisfying the property

$$K(A,x,y) = \alpha(A,x)\overline{\alpha(A,y)} \exp M(A,x,y) \tag{5.4}$$

for all $A\varepsilon\mathcal{S}$, $x,y\varepsilon X$.

<u>Proof</u>: Choose $L(A,.,.)$ satisfying (5.1) according to Theorem 5.2 and write $M(A,x,y) = L(A,x,y) - L(A,x,x_0) - L(A,x_0,y) + L(A,x_0,x_0)$ for some fixed point $x_0\varepsilon X$. $M(A,.,.)$ is positive definite by Lemma 1.7. Since $K(A,.,.)$ is continuous, its modulus is continuous, and hence the real part of $M(A,.,.)$ is continuous. By Corollary 1.4 $M(A,.,.)$ is continuous. If we write $\alpha(A,x) = \exp\{L(A,x,x_0) - \frac{1}{2}L(A,x_0,x_0)\}$, all the required properties are satisfied and the theorem is proved.

In the case of invariant kernels we shall restrict ourselves to the case where G acts transitively on X. Under this assumption it is no loss in generality to assume that G acts on itself.

<u>Theorem 5.4</u>. Let G be a complete metric group and $(\Omega, \mathcal{S})$ a standard borel space. Let further $\{\phi(A,.), A\varepsilon\mathcal{S}\}$ be a factorisable family of continuous positive definite functions on G. Then

there exists an additive family $\{\psi(A,.),\ A\epsilon\mathcal{S}\}$ of continuous conditionally positive definite functions on G such that, for every $A\epsilon\mathcal{S}$, $g\epsilon G$,

$$\exp \psi(A,g) = \phi(A,g). \tag{5.5}$$

Proof: By Theorem 5.2, there exists an additive family of conditionally positive definite functions $\{\psi(A,.),\ A\epsilon\mathcal{S}\}$ such that (5.5) holds. Without loss of generality we may assume that $\psi(A,e) = 0$ for all A. For any finite borel partition $\mathbb{P}(A) = (A_1,\ldots,A_n)$ of a set $A\epsilon\mathcal{S}$, we define a continuous function

$$\Phi(\mathbb{P}(A),.) = \sum_{k=1}^{n} |1 - \phi(A_k,.)|.$$

$\Psi(A,.) = \sup_{\mathbb{P}(A)} \Phi(\mathbb{P}(A),.)$, where the supremum is taken over all finite borel partitions of A, is then a lower semicontinuous function which is everywhere finite by Lemma 4.8. Putting, for A fixed, $D_n = \{g:\Psi(A,g)\leq n\}$, $n = 1,2,\ldots$, we get an increasing sequence of closed sets whose union is G. Since G is of second category, at least one D_{n_o} has nonempty interior. The relation

$$|\psi(A,g)| \leq \Psi(A,g),$$

which is a consequence of (5.2), implies that $|\psi(A,g)| \leq n_o$ in some open set $E\subset G$. Since $M(A,g,h) = \psi(A,h^{-1}g) - \psi(A,h^{-1}) - \psi(A,g)$ is continuous in both variables by Theorem 5.3, we may assume that $e\epsilon E$. All we have to show now is that $\psi(A,.)$ is continuous at e, because the continuity of M will then imply the continuity of $\psi(A,.)$. Since $\psi(A,e) = 0$ by our assumption, it suffices to show that $\lim_n \psi(A,g_n) = 0$ for any sequence (g_n)

converging to e. Assume there exists such a sequence (h_n) for which $\lim_n \psi(A,h_n) = c \neq 0$. From the continuity of M it follows that $\lim_n \psi(A,h_n^s) = cs$ for any integer s. Choose s_o such that $|cs_o| > n_o$. Since $(h_n^{s_o})$ still converges to e this contradicts our result that $|\psi(A,.)|$ is bounded by n_o in a neighbourhood of e. Hence $\psi(A,.)$ must be continuous. Since A was arbitrary, the theorem is proved.

§6. Representations of current groups

In this chapter we shall show that any factorisable family of projectively invariant positive definite kernels defines a projective representation of a current group.

Let Z be a set and $(\Omega, \mathfrak{S})$ a standard borel space. Let $S(\Omega,Z)$ be the space of all maps from Ω into Z taking only finitely many values and such that the inverse image of every single point set in Z lies in $\mathfrak{S}$. Then $S(\Omega,Z)$ is called the simple current space of Z over Ω. If Z is a group, then $S(\Omega,Z)$ also becomes a group under pointwise multiplication. Then it is called the simple current group of Z over Ω.
Suppose X is a set and G is a group acting on X. Then the simple current group $S(\Omega,G)$ acts on the simple current space $S(\Omega,X)$ in the following manner: $(\gamma,f)\to\gamma f$, where $\gamma f(\omega) = \gamma(\omega)f(\omega)$ for any $\gamma\in S(\Omega,G)$, $f\in S(\Omega,X)$, $\omega\in\Omega$. Suppose $\{K(A,.,.), A\in\mathfrak{S}\}$ is a factorisable family of projectively invariant positive definite kernels on X×X. We define a kernel $K^*(.,.)$ on $S(\Omega,X)\times S(\Omega,X)$

as follows: Let f_1, f_2 be two elements of $S(\Omega,X)$ such that f_1 and f_2 take values x_i and y_i respectively on A_i, where A_i, $i=1,\ldots,n$, forms a borel partition of Ω. Then we put

$$K^*(f_1,f_2) = \prod_{i=1}^{n} K(A_i,x_i,y_i). \tag{6.1}$$

It follows easily from Lemma 1.5 and the factorisability of K that K^* is a well defined positive definite kernel on $S(\Omega,X) \times S(\Omega,X)$. The projective invariance of K implies that K^* is projectively invariant under the action of the group $S(\Omega,G)$ on the space $S(\Omega,X)$.

If $\{L(A,.,.), A\in\mathcal{S}\}$ is an additive family of affine invariant conditionally positive definite kernels on $X\times X$, we define

$$L^*(f_1,f_2) = \sum_{i=1}^{n} L(A_i,x_i,y_i), \tag{6.2}$$

where f_1 and f_2 are exactly as in the preceding paragraph. It follows from Lemma 1.7 that L^* is a conditionally positive definite kernel on $S(\Omega,X)\times S(\Omega,X)$ which is affine invariant under $S(\Omega,G)$.

We shall call K^* and L^* the <u>canonical extensions</u> of K and L. Now we choose and fix a point x_o in X and call it the <u>origin</u> of X. Dividing $K(A,.,.)$ by $K(A,x_o,x_o)$ if necessary, we can always assume that $K(A,x_o,x_o) = 1$ for all A. Similarly we shall assume that $L(A,x_o,x_o) = 0$ for all A. Hereafter we shall only consider positive and conditionally positive definite kernels which take values 1 and 0 respectively at the point (x_o,x_o). Whenever X is a group the origin is always taken to be the identity element. We define for any $A\in\mathcal{S}$, $g\in G$, $x\in X$,

$$\chi_x^A(\omega) = x \text{ if } \omega \in A, \qquad = x_o \text{ if } \omega \notin A, \tag{6.3}$$

$$\chi_g^A(\omega) = g \text{ if } \omega \in A, \qquad e \text{ if } \omega \notin A. \tag{6.4}$$

Applying now Theorems 2.7 and 3.4 successively to the kernels K^* and L^* defined by (6.1) and (6.2) respectively, we obtain the following two theorems.

Theorem 6.1. Let X be a set and G be a group acting on X. Let $(\Omega,\mathcal{S})$ be a standard borel space and $\{K(A,.,.), A\epsilon\mathcal{S}\}$ be a multiplicative family of projectively invariant positive definite kernels such that

$$K(A,x_o,x_o) = 1 \text{ for all } A\epsilon\mathcal{S},$$

for some fixed point x_o in X. Let $K^*(.,.)$ be the canonical extension of $\{K(A,.,.), A\epsilon\mathcal{S}\}$ to $S(\Omega,X)\times S(\Omega,X)$, where $S(\Omega,X)$ is the simple current space of X over Ω. Then there exists a projective representation U of the current group $S(\Omega,G)$ in a Hilbert space H and a map $u:S(\Omega,X)\to H$ such that

$$\langle u(f),u(f')\rangle = K^*(f,f'), \tag{6.5}$$

$$U_\gamma u(f) = \alpha(\gamma,f)\, u(\gamma f), \tag{6.6}$$

$$\langle u(\chi_x^A),u(\chi_y^B)\rangle = K(A\cap B,x,y), \tag{6.7}$$

for all $f,f'\epsilon S(\Omega,X)$, $\gamma\epsilon S(\Omega,G)$, $A,B\epsilon\mathcal{S}$, $x,y\epsilon X$, where α is a complex valued function on $S(\Omega,G)\times S(\Omega,X)$.

<u>Theorem 6.2</u>. Let X, G and $(\Omega, \mathcal{S})$ be as in Theorem 6.1. Suppose $\{L(A,.,.),\ A\varepsilon\mathcal{S}\}$ is an additive family of affine invariant conditionally positive definite kernels on $X\times X$ such that

$$L(A,x_o,x_o) = 0 \text{ for all } A\varepsilon\mathcal{S},$$

for some fixed point x_o in X. Let L^* be the canonical extension of $\{L(A,.,.),\ A\varepsilon\mathcal{S}\}$ to $S(\Omega,X)\times S(\Omega,X)$ where $S(\Omega,X)$ is the simple current space of X over Ω. Let $f_o\varepsilon S(\Omega,X)$ be the function which takes value x_o everywhere on Ω. Then there exists a representation V of the current group $S(\Omega,G)$ in a Hilbert space H' and a first order cocycle v on $S(\Omega,X)$ with origin f_o and values in H' such that

$$\langle v(f),v(f')\rangle = L^*(f,f') - L^*(f,f_o) - L^*(f_o,f'), \qquad (6.8)$$

$$V_\gamma v(f) = v(\gamma f) - v(\gamma f_o), \qquad (6.9)$$

$$\langle v(\chi_x^A),v(\chi_y^B)\rangle = L(A\cap B,x,y) - L(A\cap B,x,x_o) - L(A\cap B,x_o,y), \quad (6.10)$$

for all $f,f'\varepsilon S(\Omega,X)$, $\gamma\varepsilon S(\Omega,G)$, $A,B\varepsilon\mathcal{S}$ and $x,y\varepsilon X$.

<u>Remark 6.3</u>. In order that Theorems 6.1 and 6.2 may be valid it is enough to assume that $K(A,.,.)$ and $L(A,.,.)$ are finitely multiplicative and finitely additive in A.

<u>Remark 6.4</u>. If we assume $\{K(A,.,.)\}$ to be a factorisable family of invariant positive definite kernels and $\{L(A,.,.)\}$ to be an additive family of invariant kernels, then both the kernels K^* and L^* will be invariant under the current group. Hence the representation $\gamma\to U_\gamma$ of the current group $S(\Omega,G)$ will be an ordinary representation.

Remark 6.5. According to Theorem 5.2 every projectively invariant factorisable family $\{K(A,.,.)\}$ of positive definite kernels has the property that there exists a family $\{L(A,.,.)\}$ of affine invariant conditionally positive definite kernels with $K(A,x,y) = \exp L(A,x,y)$ for all $A \in \mathcal{S}'$, $x,y \in X$. By Theorem 6.1, K gives rise to a projective representation $\gamma \to U_\gamma$ of the simple current group $S(\Omega,G)$ satisfying (6.5) - (6.7). By Theorem 6.2, L gives rise to an ordinary representation $\gamma \to V_\gamma$ of $S(\Omega,G)$ and a cocycle v. What is the relation between U,V and v? We shall try to answer this question in the next section. The Araki-Woods imbedding theorem will follow from this analysis as a apecial case.

§7. Continuous tensor products of Hilbert spaces and the Araki-Woods imbedding theorem

The main aim of this section is to formulate the notion of a continuous tensor product of Hilbert spaces and describe continuous tensor products of representations of a group.
Let H_1 and H_2 be Hilbert spaces. We define the following kernel F on the cartesian product $H_1 \times H_2$:

$$F((u,v),(u',v')) = \langle u,u' \rangle \langle v,v' \rangle,$$

for all $u,u' \in H_1$, $v,v' \in H_2$, where $\langle .,. \rangle$ denotes the inner product in the corresponding space. Since the kernels F_i, $i=1,2$, on $H_1 \times H_2$ given by

$$F_1((u,v),(u',v')) = \langle u,u' \rangle, \quad F_2((u,v),(u',v')) = \langle v,v' \rangle,$$

are both positive definite, Lemma 1.7 implies that their product F is also a positive definite kernel on $H_1 \times H_2$. Let J be the free linear space over $H_1 \times H_2$. For any two elements $w_1 = \sum_{j=1}^{n} a_j\ (u_j, v_j)$ and $w_2 = \sum_{j=1}^{m} b_j\ (u_j', v_j')$ we define a form $\langle .,. \rangle$ by

$$\langle w_1, w_2 \rangle = \sum_{j=1}^{n} \sum_{k=1}^{m} a_j \overline{b_k}\ F((u_j, v_j), (u_k', v_k')).$$

It is easy to see that the set of all elements w in J for which $\langle w, w \rangle = 0$ forms a linear subspace J_o of J, and that the form $\langle .,. \rangle$ defines an inner product on the quotient space J/J_o. We denote the completion of J/J_o with respect to this inner product, for which we still write $\langle .,. \rangle$, by $H_1 \otimes H_2$ and we call this space the <u>tensor product</u> of the two spaces H_1 and H_2. We shall write $u \otimes v$ for the image of the element (u,v) under the map from $H_1 \times H_2$ into $H_1 \otimes H_2$ for all $u \in H_1$, $v \in H_2$. If (u_i) is an orthonormal basis in H_1 and (v_i) is an orthonormal basis in H_2, then the set of vectors $(u_i \otimes v_j)$, $i,j = 1,2,\ldots$ forms an orthonormal basis in $H_1 \otimes H_2$.

Suppose $H_i = L_2(X_i, \mathcal{S}_i, \mu_i)$, $i=1,2$, where $(X_i, \mathcal{S}_i, \mu_i)$, $i=1,2$, are two measure spaces. If $f \in H_1$ and $g \in H_2$, then the function $f(x)g(y)$ of two variables $x \in X_1$, $y \in X_2$ is square integrable with respect to $\mu_1 \times \mu_2$ and hence belongs to $H = L_2(X_1 \times X_2, \mathcal{S}_1 \times \mathcal{S}_2, \mu_1 \times \mu_2)$. It is easy to see that there is a unitary map from $H_1 \otimes H_2$ onto H which takes the element $f \otimes g$ into the function $f(x)g(y)$.

Consider the case $H_1 = H_2 = H$. The closed subspace generated by the elements of the form $u \otimes v + v \otimes u$ in $H \otimes H$ is called the symmetric tensor product of H with itself and denoted by $H \circledS H$. By an obvious generalisation of the above we define for any n

Hilbert spaces $H_1,\ldots,H_n$ their tensor product $H_1\otimes\ldots\otimes H_n$, which is generated by all elements of the form $u_1\otimes\ldots\otimes u_n$, $u_i\varepsilon H_i$, and which is separated and completed with respect to the inner product defined by $\langle u_1\otimes\ldots\otimes u_n, v_1\otimes\ldots\otimes v_n\rangle = \prod_{i=1}^{n}\langle u_i,v_i\rangle$, where $u_i,v_i\varepsilon H_i$ for $i=1,\ldots,n$. Since every Hilbert space is unitarily isomorphic to a space of square integrable functions on some measure space, it will be sufficient to explain the notion of an n-fold symmetric tensor product for L_2-spaces. Let $(X,\mathcal{S},\mu)$ be a measure space. As before, we have a natural unitary mapping from $L_2(X\times\ldots\times X,\mathcal{S}\times\ldots\times\mathcal{S},\mu\times\ldots\times\mu)$ onto $L_2(X,\mathcal{S},\mu)\otimes\ldots\otimes L_2(X,\mathcal{S},\mu)$, where each factor is taken n times. Each permutation of the n variables in $L_2(X\times\ldots\times X,\mathcal{S}\times\ldots\times\mathcal{S},\mu\times\ldots\times\mu)$ induces therefore a unitary transformation in $H = L_2(X,\mathcal{S},\mu)\otimes\ldots\otimes L_2(X,\mathcal{S},\mu)$ (n times). We denote the closed subspace of all vectors in H which are invariant under <u>all</u> of these unitary transformations by $L_2(X,\mathcal{S},\mu)\,ⓢ\ldots ⓢ\,L_2(X,\mathcal{S},\mu)$ and call it the n-fold symmetric tensor product of $L_2(X,\mathcal{S},\mu)$. It can be identified with the space of symmetric functions in n variables which are square integrable with respect to the product measure $\mu\times\ldots\times\mu$.

If T_1 and T_2 are bounded linear operators defined on Hilbert spaces H_1 and H_2 respectively and taking values in H_1' and H_2', then there is a unique bounded linear operator defined on $H_1\otimes H_2$ and taking values in $H_1'\otimes H_2'$, which maps the element $u\otimes v$ into $T_1u\otimes T_2v$ for all $u\varepsilon H_1$, $v\varepsilon H_2$. We denote this operator by $T_1\otimes T_2$. If T_1 and T_2 is unitary, then $T_1\otimes T_2$ is also unitary.

The countable direct sum $C\oplus H\oplus H\otimes H + \ldots$ is called the <u>Fock space</u> of H. The countable direct sum $C\oplus H\oplus H\,ⓢ\,H\oplus H\,ⓢ\,H\,ⓢ\,H+\ldots$

is called the symmetric Fock space over H and is denoted by Exp H. Here C stands for the complex numbers considered as a one dimensional Hilbert space. For every $v \in H$, we write

$$\text{Exp } v = 1 \oplus v \oplus \frac{1}{\sqrt{2}} v \otimes v \oplus \frac{1}{\sqrt{3!}} v \otimes v \otimes v \oplus \ldots \qquad (7.1)$$

Clearly Exp v ε Exp H. We have

$$\langle \text{Exp } v, \text{Exp } v' \rangle = \exp \langle v, v' \rangle, \quad v, v' \in H. \qquad (7.2)$$

It is not difficult to verify that the vectors $\{\text{Exp } v,\ v \in H\}$ span Exp H. If H_1 and H_2 are two Hilbert spaces, then $\text{Exp}\ (H_1 \oplus H_2)$ can be mapped unitarily onto $\text{Exp } H_1 \otimes \text{Exp } H_2$ in such a manner that $\text{Exp}\ (u \oplus v)$ goes to $\text{Exp } u \otimes \text{Exp } v$ for all $u \in H_1$, $v \in H_2$.

Let $(\Omega, \mathcal{S}, \mu)$ be a measure space and $\{H^{\omega}, \omega \in \Omega\}$ a measurable family of Hilbert spaces. Put H equal to the direct integral $\int_{\Omega} H^{\omega}\, d\mu(\omega)$ and $H(A) = \int_{A} H^{\omega}\, d\mu(\omega)$ for any $A \in \mathcal{S}$. If A and B are disjoint sets in $\mathcal{S}$ it follows that $H(A \cup B) = H(A) \oplus H(B)$ and hence there is a unitary map $U(A,B)$ from $\text{Exp } H(A) \otimes \text{Exp } H(B)$ onto $\text{Exp } H(A \cup B)$ defined by

$$U(A,B)\ (\text{Exp } u \otimes \text{Exp } v) = \text{Exp}\ (u \oplus v)$$

for all $u \in H(A)$, $v \in H(B)$. Let $I(A)$ denote the identity operator in Exp H(A). It is easy to see that, for any three disjoint borel sets A,B,C,

$$U(A \cup B, C)\{U(A,B) \otimes I(C)\} = U(A, B \cup C)\{I(A) \otimes U(B,C)\}. \qquad (7.3)$$

This suggests the following definition:

<u>Definition 7.1</u>. Let $(\Omega,\mathcal{S})$ be a borel space. Let $\{\mathcal{H}(A), A\in\mathcal{S}\}$ be a family of Hilbert spaces satisfying the following: for any two disjoint borel sets A,B there is a unitary map U(A,B) from $\mathcal{H}(A)\otimes\mathcal{H}(B)$ onto $\mathcal{H}(A\cup B)$ such that (7.3) is satisfied for any three disjoint borel sets A,B and C. Then the family is said to be <u>factorisable</u>. If $(\Omega,\mathcal{S})$ is standard and $\mathcal{H}(\{\omega\})$ is one dimensional for every single point set $\{\omega\},\omega\in\Omega$, then $\mathcal{H}(\Omega)$ is said to be a <u>continuous tensor product</u>.

It is clear that if μ is a nonatomic measure on a standard borel space $(\Omega,\mathcal{S})$ and $\mathcal{H}(A) = \text{Exp } H(A)$, where $H(A) = \int_A H^{\omega}\, d\mu(\omega)$ and $\{H^{\omega},\omega\in\Omega\}$ is a measurable family of Hilbert spaces on Ω, then $\mathcal{H}(\Omega)$ is a continuous tensor product. Let $H(\Omega) = H$ and let P(A) be the prjection on the subspace H(A). For any $A\in\mathcal{S}$, $u,v\in H$, we define

$$K(A,u,v) = \exp \langle P(A)u,v\rangle .$$

Then it is obvious that $\{K(A,.,.), A\in\mathcal{S}\}$ is a factorisable family of positive definite kernels on $H\times H$. Now we raise the question whether factorisable families of positive definite kernels give rise to continuous tensor products of Hilbert spaces in some special manner. That this is so will be clear from our subsequent analysis. However , the problem of finding the most general continuous tensor product is still open.

<u>Theorem 7.2</u>. Let X be a set and G be a group acting on it. Let L(.,.) be an affine invariant conditionally positive definite

kernel such that

$$L(x_o, x_o) = 0 \text{ for some fixed point } x_o \varepsilon X,$$

$$L(gx, gy) = L(x,y) + \beta(g,x) + \overline{\beta(g,y)},$$

for all $x, y \varepsilon X$, $g \varepsilon G$. Suppose

$$K(x,y) = \exp L(x,y) \text{ for all } x,y \varepsilon X.$$

Then $K(.,.)$ is a projectively invariant positive definite kernel on $X \times X$. Further there exists a unitary representation V of G in a Hilbert space H, a first order cocycle v on X with origin x_o and values in H, a subspace $\mathcal{H}$ of Exp H, a projective representation U of G in $\mathcal{H}$ and a map u from X into $\mathcal{H}$ so that the following properties hold for all $g \varepsilon G$, $x, y \varepsilon X$:

(1) $L(x,y) - L(x,x_o) - L(x_o,y) = \langle v(x), v(y) \rangle$,

(2) $V_g v(x) = v(gx) - v(gx_o)$,

(3) $u(x) = \{\exp L(x,x_o)\}. \operatorname{Exp} v(x)$,

(4) $\langle u(x), u(y) \rangle = K(x,y)$

(5) $U_g u(x) = \{\exp -\beta(g,x)\}.u(gx)$

(6) $\mathcal{H}$ is spanned by $\{u(x), x \varepsilon X\}$.

<u>Proof</u>: That K is a projectively invariant positive definite kernel follows from Lemma 1.7. Properties (1) and (2) constitute a restatement of Theorem 3.4. Property (3) is the definition of the map u. Property (6) is the definition of the subspace $\mathcal{H}$ of the symmetric Fock space Exp H. Propert (4) follows from (7.2). Property (5) is a consequence of Theorem 2.7.

Remark 7.3. The projective representation U of the above theorem satisfies

$$U_{g_1}U_{g_2} = \exp i\, s(g_1,g_2)\, U_{g_1g_2} \quad \text{for all } g_1,g_2 \in G,$$

where s(.,.) is a second order cocycle of G with values in the real line, satisfying the identity

$$\beta(g_1g_2,x) - \beta(g_1,g_2x) - \beta(g_2,x) = i\, s(g_1,g_2)$$

for all $g_1,g_2 \in G$, $x \in X$. In fact this is a consequence of Lemma 3.2.

Remark 7.4. If G is a topological group acting continuously on a topological space X and L(.,.) is a continuous affine invariant conditionally positive definite kernel on X×X, then the maps v and u of the theorem are continuous and the representations V and U of G are weakly continuous.

Theorem 7.5. Let G be a group acting on a set X and $(\Omega,\mathcal{S})$ be a standard borel space. Suppose that $\{L(A,.,.),\ A\in\mathcal{S}\}$ is an additive family of affine invariant conditionally positive definite kernels such that $L(A,x_o,x_o) = 0$ for some fixed point $x_o \in X$ and all $A \in \mathcal{S}$. Then there exists a unitary representation V of G in a Hilbert space H, a first order cocycle v on X with origin x_o and values in H and a projection valued measure P defined on $\mathcal{S}$ and operating in H such that the following properties hold:

$$V_gP(A) = P(A)V_g \text{ for all } g\in G,\ A\in\mathcal{S},$$

$$L(A,x,y) - L(A,x,x_o) - L(A,x_o,y) = \langle P(A)v(x),v(y)\rangle \quad \text{for all } A\varepsilon\mathcal{S},\ x,y\varepsilon X.$$

Conversely if V is a unitary representation of G in H, v a first order cocycle on X with values in H and origin x_o and P is a projection valued measure on $\mathcal{S}$ operating in H and commuting with V, then for any complex valued function $\beta(A,x)$ which is countably additive in A for every fixed x and identically zero at x_o, the kernels $L(A,.,.)$ defined by

$$L(A,x,y) = \langle P(A)v(x),v(y)\rangle + \beta(A,x) + \overline{\beta(A,y)}$$

constitute an additive family of affine invariant conditionally positive definite kernels.

Proof: Let L^* be the canonical extension of $\{L(A,.,.),\ A\varepsilon\mathcal{S}\}$ to $S(\Omega,X)\times S(\Omega,X)$, where $S(\Omega,X)$ is the simple current space of X over Ω. Then by Theorem 6.2 there exists a representation V of the simple current group $S(\Omega,G)$ in a Hilbert space H and a first order cocycle v on $S(\Omega,X)$ with values in H such that equations (6.8) - (6.10) are fulfilled. Let $A_1,\ldots,A_k,B_1,\ldots,B_l$ together form a borel partition of Ω. Let f_1,f_2 be elements in $S(\Omega,X)$ where f_1 takes the values $x_1,\ldots,x_k$ on $A_1,\ldots,A_k$ respectively and the value x_o on all the B_i, and f_2 takes value x_o on all the A_j and values $y_1,\ldots,y_l$ on $B_1,\ldots,B_l$ respectively. Whenever this happens we shall say that f_1 and f_2 have disjoint supports contained in $\cup A_j$ and $\cup B_j$ respectively. Let f be the function which takes values x_j on the sets A_j, j=1...k, and y_j on B_j, j=1...l. Then it is not difficult to verify that for any

$f' \in S(\Omega, X)$,

$$L^*(f,f') - L^*(f,f_o) - L^*(f_o,f') =$$
$$= L^*(f_1,f') - L^*(f_1,f_o) - L^*(f_o,f') + L^*(f_2,f') - L^*(f_2,f_o) - L^*(f_o,f').$$

Hence, from (6.8),

$$\langle v(f),v(f')\rangle = \langle v(f_1),v(f')\rangle + \langle v(f_2),v(f')\rangle$$

for all $f' \in S(\Omega,X)$. Thus

$$v(f) = v(f_1) + v(f_2) \tag{7.4}$$

whenever f, f_1, f_2 are as described earlier. We can define f as the 'product' of f_1 and f_2. Of course, this is meaningful only when f_1 and f_2 have disjoint support. It follows from (6.8) and (6.9) that

$$\langle v(f_1),v(f_2)\rangle = 0 \tag{7.5}$$

whenever f_1 and f_2 have disjoint supports. For any $A \in \mathcal{S}$, let H(A) be the closed linear span of all the vectors $\{v(f'), f' \in S(\Omega,X)$, support $f' \subset A\}$. Then (7.5) implies that H(A) and H(B) are orthogonal whenever A and B are disjoint. By (7.4), H(A) is spanned by all vectors of the form $v(\chi_x^E)$, $E \subset A$, $x \in X$, where χ_x^E is defined by (6.3). It follows from (7.4) and the countable additivity of L(A,x,y) in the variable A, that

$$v(\chi_x^{\cup E_i}) = \sum_{i=1}^{\infty} v(\chi_x^{E_i})$$

whenever $E_1, E_2, \ldots$ are disjoint borel sets and $x \in X$.

This shows that

$$H(\cup E_i) = \bigoplus_{i=1}^{\infty} H(E_i)$$

for any sequence E_i of disjoint borel sets. If we denote the projection onto the subspace $H(E)$ by $P(E)$, then $P(.)$ is a projection valued measure on $\mathcal{S}$. We now write

$$\begin{aligned} v(E,x) &= v(\chi_x^E), \\ v(x) &= v(\Omega,x), \qquad (7.6) \\ V_g &= V_{\chi_g^\Omega}, \end{aligned}$$

for all $E \in \mathcal{S}$, $x \in X$, $g \in G$. It follows from (7.4) and (7.5) that

$$P(E)v(F,x) = v(E \cap F,x). \qquad (7.7)$$

Further the vectors $v(F,x)$ span the whole space H as F varies over $\mathcal{S}$ and x over X. Now (6.10) and (7.7) imply that

$$\langle P(E)v(x),v(y)\rangle = L(E,x,y) - L(E,x,x_o) - L(E,x_o,y) \qquad (7.8)$$
$$\text{for all } E \in \mathcal{S},\ x,y \in X.$$

Since $v(f)$, $f \in S(\Omega,X)$ is a cocycle with origin f_o for the representation V_γ it follows that $v(x)$ is a cocycle for the representation V of G. From the cocycle equation (6.9) and (7.7) we obtain

$$\begin{aligned} V_g P(E)v(F,x) &= P(E)\, V_g v(F,x) \\ &= v(E \cap F,gx) - v(E \cap F,gx_o). \end{aligned}$$

This equation together with (7.8) completes the proof of the first part. The converse is verified in a straightforward manner. This completes the proof of the theorem.

<u>Remark 7.6</u>. If G is a topological group which acts continuously on a topological space X, and if $L(A,.,.)$ is continuous for every A, then the maps $g \to V_g$ and $x \to v(x)$ are continuous.

As in the proof of Theorem 7.5 we shall say that an element $f \in S(\Omega,X)$ vanishes at a point ω if $f(\omega) = x_o$. If f vanishes at all points in the complement of a set $A \in \mathcal{S}$, we shall say that the support of f is contained in A. If G is a group, we shall say that $\gamma \in S(\Omega,G)$ vanishes at a point ω if $\gamma(\omega) = e$, the identity element of G. If γ vanishes at all points in the complement of a set $A \in \mathcal{S}$, we say that the support of γ is contained in A. We write

$$S(A,X) = \{f: \text{support } f \subset A,\ f \in S(\Omega,X)\},$$
$$S(A,G) = \{\gamma: \text{support } \gamma \subset A,\ \gamma \in S(\Omega,G)\}$$

for every $A \in \mathcal{S}$. It is clear that $S(A,G)$ is a subgroup of $S(\Omega,G)$. Further, for any two disjoint borel sets A and B , $S(A \cup B,G)$ is the cartesian product $S(A,G) \times S(B,G)$ of the subgroups $S(A,G)$ and $S(B,G)$.

<u>Theorem 7.7</u>. Let G be a group acting on a set X and let $(\Omega,\mathcal{S})$ be a standard borel space. Suppose V is a unitary representation of G in a Hilbert space H, v is a first order cocycle on X with origin x_o and values in H, and P is a projection valued measure on $\mathcal{S}$, acting on H and commuting with V. For any $f \in S(\Omega,X)$ which takes values $x_1,\ldots,x_k$ on disjoint borel sets $A_1,\ldots,A_k$, whose union is Ω, let

$$v(f) = \sum_{i=1}^{k} P(A_i)v(x_i).$$

For any $\gamma \varepsilon S(\Omega,G)$ which takes values $g_1,\ldots,g_l$ on the elements $B_1,\ldots,B_l$ of a borel partition of Ω, let

$$V_\gamma = \sum_{j=1}^{l} P(B_j)V_{g_j}.$$

Then $\gamma \to V_\gamma$ is a unitary representation of $S(\Omega,G)$ in H and $f \to v(f)$ is a first order cocycle in $S(\Omega,X)$ with origin f_o and values in H, where f_o is identically equal to x_o on Ω.

Let $\mathcal{H}(A) \subset \text{Exp } H$ be the subspace spanned by all the vectors $\{\text{Exp } v(f), \text{ support } f \subset A\}$, and let $\mathcal{H} = \mathcal{H}(\Omega)$. Then the family of Hilbert spaces $\{\mathcal{H}(A), A\varepsilon\mathcal{S}\}$ is factorisable. If P is a non-atomic projection valued measure, then $\mathcal{H}$ is a continuous tensor product.

The function L^*, defined on $S(\Omega,X)\times S(\Omega,X)$ by

$$L^*(f_1,f_2) = \langle v(f_1),v(f_2)\rangle,$$

is an affine invariant conditionally positive definite kernel under the action of $S(\Omega,G)$. Suppose

$$L^*(\gamma f_1,\gamma f_2) = \beta(\gamma,f_1) + \overline{\beta(\gamma,f_2)} + L^*(f_1,f_2) \quad \text{for all } f_1,f_2 \varepsilon S(\Omega,X),\ \gamma\varepsilon S(\Omega,G).$$

then there is a unique projective representation $\gamma \to U_\gamma$ in of $S(\Omega,G)$ such that

$$U_\gamma \text{Exp } v(f) = \{\exp -\beta(\gamma,f)\}.\text{Exp } v(\gamma f) \quad \text{for all } f\varepsilon S(\Omega,X).$$

U_γ leaves $\mathcal{H}(A)$ invariant whenever the support of γ is contained in A and hence defines a projective representation U^A of the subgroup S(A,G) in $\mathcal{H}(A)$. Under the correspondence between $\mathcal{H}(A) \otimes \mathcal{H}(B)$ and $\mathcal{H}(A \cup B)$ for disjoint borel sets A and B, the representations $U^A \otimes U^B$ and $U^{A \cup B}$ for the subgroup $S(A \cup B,G)$ correspond.

Proof: The first part of the theorem follows from straightforward verification. For proving the second part, let A and B be disjoint borel sets. Define the map U(A,B) from $\mathcal{H}(A) \otimes \mathcal{H}(B)$ onto $\mathcal{H}(A \cup B)$ by extending the map

$$U(A,B)\{\mathrm{Exp}\, v(f_1) \otimes \mathrm{Exp}\, v(f_2)\} = \mathrm{Exp}\, v(f) \qquad (7.9)$$

for any f_1 with support in A and f_2 with support in B, where f is the element which agrees with f_1 inside A and with f_2 outside A. Since $v(f) = v(f_1) + v(f_2)$ and $v(f_1)$ and $v(f_2)$ are orthogonal, it is clear that U(A,B) is a unitary isomorphism. That U(A,B) satisfies equation (7.3) follows from the discussion before Definition 7.1. This proves the second part. That the map $\gamma \to U_\gamma$ defines a projective representation in $\mathcal{H}$ follows from Theorem 7.2. From the cocycle equation (3.2) for v(f), it follows that

$$\beta(\gamma,f) = \langle v(\gamma f), v(\gamma f_0)\rangle - \tfrac{1}{2}\langle v(\gamma f_0), v(\gamma f_0)\rangle .$$

If A and B are disjoint borel sets, γ_1 and f_1 have support in A and γ_2 and f_2 have support in B, $\gamma = \gamma_1\gamma_2$ and f is the function which coincides with f_1 inside A and with f_2 outside A, then

$$\beta(\gamma_1,f_1) + \beta(\gamma_2,f_2) = \beta(\gamma,f).$$

This together (7.9) and the definition of U and U(A,B) imply that

$$U(A,B)\, U^A_{\gamma_1} \otimes U^B_{\gamma_2} = U^{A\cup B}_{\gamma_1\gamma_2}\, U(A,B).$$

This completes the proof of the theorem.

Remark 7.8. Theorems 5.2, 7.5 and 7.7 clearly indicate how factorisable families of projectively invariant positive definite kernels give rise to factorisable families of Hilbert spaces and a projective representation of the current group with a factorisability property. Further this factorisable family of Hilbert spaces can be imbedded in a single symmetric Fock space. This is a generalisation of the Araki-Woods imbedding theorem. Araki calls the representations of current groups satisfying the last property of Theorem 7.7 factorisable representations.

Remark 7.9. We introduce the smallest topology in $S(\Omega,G)$ which makes the map $\gamma \to v(\gamma f)$ and $\gamma \to v(\gamma^{-1} f)$ continuous for every $f \in S(\Omega,X)$. In this topology, the representation U of Theorem 7.7 is continuous. Obviously, the representation can be extended to the completion of the group $S(\Omega,G)$ in this topology. Note that this topology is induced by a family of pseudometrics.

The continuous tensor product constructed in the Theorems 7.5 and 7.7 is actually a symmetric Fock space over a direct integral of Hilbert spaces (this follows from a straightforward application of the Hahn-Hellinger theorem to the projection valued measure P). However, in general it does not seem possible to obtain the representation V as a direct integral of representations. We shall now make the additional assumption that the group G is locally compact and acts on itself by left multiplication. Under these circumstances we will show that both the represantation V and the cocycle v can be understood as direct integrals.

<u>Theorem 7.10</u>. Let G be a locally compact second countable group and $(\Omega,\mathcal{S})$ a standard borel space. Suppose V is a continuous unitary representation of V in a Hilbert space H, v is a continuous first order cocycle on G with values in H and origin e, where e is the identity of G. Let further P be a projection valued measure on $\mathcal{S}$, which operates in H and commutes with V. Assume finally that H is generated by the set $\{v(g), g\in G\}$. Then there exists a measurable family $\{H^{\omega}, \omega\in\Omega\}$ of Hilbert spaces, a totally finite measure μ on $\mathcal{S}$, families $\{V^{\omega}, \omega\in\Omega\}$ and $\{v(\omega,.), \omega\in\Omega\}$ of weakly continuous unitary representations of G in H^{ω} and of continuous cocycles on G with values in H^{ω} and origins e, respectively, such that

(1) $V_g^{\omega}\, v(\omega,h) = v(\omega,gh) - v(\omega,g)$ for all $\omega\in\Omega$, $g,h\in G$,

(2) $V_g = \int_{\Omega} V_g^{\omega}\, d\mu(\omega)$ for all $g\in G$,

(3) $v(g) = \int_{\Omega} v(\omega,g)\, d\mu(\omega)$ for all $g\in G$,

(4) $P(A) = \int_{A} I^{\omega}\, d\mu(\omega)$ for any $A\in\mathcal{S}$,

where I^{ω} stands for the identity operator in H^{ω}. The measure μ is unique up to equivalence, and the map $\omega \to (H^{\omega}, V^{\omega}, v(\omega,.))$ is unique up to unitary equivalence in each component almost everywhere $\omega(\mu)$.

Proof: Since G is separable and v is continuous, H is separable. Applying the Hahn-Hellinger theorem to H and the projection valued measure P, we can decompose Ω into a countable number of disjoint borel sets B_n, $n = \aleph_0, 1, 2, \ldots$, and H into a direct sum of direct integrals

$$H = \bigoplus_{n=\aleph_0,1,2,\ldots} \int_{B_n} H_n \, d\nu_n, \qquad (7.10)$$

where, for each n, H_n is the complex Hilbert space of dimension n and ν_n is a measure whose support is contained in B_n, such that $\mu = \sum_n \nu_n$ is totally finite. If we put $H^{\omega} = H_n$ whenever $\omega \varepsilon B_n$, we may rewrite (7.10) in the form $H = \int_{\Omega} H^{\omega} \, d\mu(\omega)$ and put $P(A) = \int_A I^{\omega} \, d\mu(\omega)$, where I^{ω} is the identity operator on H^{ω}. For any $g \varepsilon G$, $A \varepsilon \mathcal{S}$, V_g and P(A) commute. A standard application of a result by von Neumann and Fubini's theorem allows it to assume that V_g is indeed a direct integral $\int_{\Omega} V_g^{\omega} \, d\mu(\omega)$, where each V^{ω} is a continuous unitary representation of G in H^{ω}. Applying now Theorem 13.2 to the representation V we find that we can indeed write

$$v(g) = \int_{\Omega} v(\omega, g) \, d\mu(\omega)$$

for any $g \varepsilon G$, where each $v(\omega,.)$ is a continuous cocycle with origin e and values in H^{ω}. The uniqueness of μ up to equivalence is a direct consequence of the Hahn-Hellinger theorem, and the other uniqueness assertions are obvious. This proves the theorem.

Remark 7.11. Theorem 7.10 has some interesting consequences concerning the existence of 'density functions' for additive families of continuous conditionally positive definite kernels. Let G be a locally compact second countable group and let $\{L(A,.,.), A\in\mathcal{S}\}$ be an additive family of continuous affine invariant conditionally positive definite kernels on G×G, such that $L(A,e,e) = 0$ for all A, where e is the identity of G. It follows from Theorem 7.5, Remark 7.6 and Theorem 7.10 that there exists a family $\{M(\omega,.,.), \omega\in\Omega\}$ of continuous affine invariant positive definite kernels on G×G and a measure μ on $\mathcal{S}$, such that

(1) $M(.,g.h)$ is μ-measurable for every $g,h\in G$,

$$\text{(2)} \int_A M(\omega,g,h)\, d\mu(\omega) = M(A,g,h) \text{ for every } A\in\mathcal{S},\ g,h\in G, \tag{7.11}$$

where $M(A,g,h)$ is defined by

$$M(A,g,h) = L(A,g,h) - L(A,g,e) - L(A,e,h).$$

To see that, it suffices to put in Theorem 7.10

$$M(\omega,g,h) = \langle v(\omega,g),v(\omega,h)\rangle \tag{7.12}$$

for every $\omega\in\Omega$, $g,h\in G$. If we assume that all kernels $L(A,.,.)$ are invariant, then we get a further result:

Theorem 7.12. Let G be a locally compact second countable group and $(\Omega,\mathcal{S})$ a standard borel space. Suppose $\{\psi(A,.), A\in\mathcal{S}\}$ is an additive family of continuous normalised conditionally positive definite functions on G. Then there exists an additive family of continuous realvalued homomorphisms $\{\xi(A,.),\ A\in\mathfrak{S}\}$ of G, a totally finite measure μ on $\mathfrak{S}$, and a family $\{\psi(\omega,.),\ \omega\in\Omega\}$ of continuous normalised conditionally positive definite functions on G, such that

(1) $\psi(\circ,g)$ is μ-measurable for every $g\in G$,

(2) $\int_A \psi(\omega,g)\, d\mu(\omega) = \psi(A,g) - i\,\xi(A,g)$ for every $A\in\mathfrak{S}$, $g\in G$.

<u>Proof</u>: In the notation of Theorem 7.5, 7.10 and Remark 7.11 we put

$$\langle P(A)v(g),v(h)\rangle = M(A,g,h) \qquad (7.13)$$

$$= \psi(A,h^{-1}g) - \psi(A,h^{-1}) - \psi(A,g).$$

(7.11) and (7.12) imply

$$\mathrm{Re}\int_A \langle v(\omega,g),v(\omega,h)\rangle\, d\mu(\omega) = \mathrm{Re}\psi(A,h^{-1}g)-\mathrm{Re}\psi(A,h^{-1})-\mathrm{Re}\psi(A,g).$$

Since each $v(\omega,.)$ is a cocycle of first order, we have, for every ω, g,

$$v(\omega,g^{-1}) = -V^{\omega}_{g^{-1}}\, v(\omega,g)$$

and hence

$$||v(\omega,gh)||^2 = ||\, v(\omega,g^{-1}) - v(\omega,h)||^2 =$$

$$= ||v(\omega,g)||^2 + ||v(\omega,h)||^2 - 2\,\mathrm{Re}\langle v(\omega,h),v(\omega,g^{-1})\rangle. \qquad (7.14)$$

This implies that for every $A\in\mathfrak{S}$, $g\in G$,

$$\mathrm{Re}\,\psi(A,g) = -\tfrac{1}{2}\int_A ||v(\omega,g)||^2\, d\mu(\omega).$$

We now write

$$\mathrm{Im}\,\psi(A,g) = \xi(A,g) + \eta(A,g),$$

where for each $g\varepsilon G$, $\xi(.,g)$ and $\eta(.,g)$ denote the singular and the absolutely continuous part of $\mathrm{Im}\,\psi(.,g)$ with respect to μ. For any two fixed elements $g_1,g_2\varepsilon G$, let $D\varepsilon\mathcal{S}$ be a set for which

$$\xi(B,g_1) = \xi(B,g_2) = \xi(B,g_1g_2) = 0$$

for all $B\subset D$, and $\mu(D') = 0$, where D' denotes the complement of D. Then

$$\xi(A,g_1\, g_2) - \xi(A,g_1) - \xi(A,g_2) =$$

$$= \mathrm{Im}\,\psi(A\wedge D',g_1g_2) - \mathrm{Im}\,\psi(A\cap D',g_1) - \mathrm{Im}\,\psi(A\cap D',g_2)$$

$$= \mathrm{Im}\int_{A\cap D'} \langle v(\omega,g_2),v(\omega,g_1^{-1})\rangle\, d\mu(\omega) = 0$$

In other words, $\xi(A,.)$ is a homomorphism from G into the additive group of real numbers for every $A\varepsilon\mathcal{S}$. Since $\eta(A,.)$ is measurable for every $A\varepsilon\mathcal{S}$, it follows that $\xi(A,.)$ is measurable and hence continuous. We now write, for every $A\varepsilon\mathcal{S}$, $g\varepsilon G$,

$$\Psi(A,g) = \psi(A,g) - i\,\xi(A,g).$$

By construction, $\Psi(.,g)$ is absolutely continuous with respect to μ for every g. Let $f(.,g)$ be the Radon Nikodym derivative

$$f(.,g) = \frac{d\Psi(.,g)}{d\mu(.)}.$$

Since $\xi(A,.)$ is a homomorphism, we have for every $g,h\epsilon G$,

$$<v(\omega,g),v(\omega,h^{-1})> = f(\omega,hg) - f(\omega,h) - f(\omega,g)$$

a.e $\omega(\mu)$. Applying Fubini's theorem to the function

$$|<v(\omega,g),v(\omega,h^{-1})> - f(\omega,hg) + f(\omega,h) + f(\omega,g)|,$$

we conclude that there exists a set $A\epsilon\mathfrak{S}$ with $\mu(A') = 0$, such that for all $\omega\epsilon A$,

$$<v(\omega,g),v(\omega,h^{-1})>= f(\omega,hg) - f(\omega,h) - f(\omega,g) \qquad (7.15)$$

a.e.g,h. This implies

$$f(\omega,g) = <v(\omega,g),v(\omega,h^{-1})> - f(\omega,hg) + f(\omega,h).$$

Let a be a continuous realvalued fuction on G vanishing outside a compact set so that $\int a(g)\, dg = 1$, where dg denotes integration with respect to the right invariant Haar measure. Then we have

$$f(\omega,g) = \int a(h)\{<v(\omega,g),v(\omega,h^{-1})> - f(\omega,hg) + f(\omega,h)\}\, dh$$

a.e.g for all $\omega\epsilon A$. Since the right hand side of this equation is continuous in g we may assume that $f(\omega,.)$ is continuous, even for all $\omega\epsilon\Omega$. Then (7.15) holds for all ω,g,h. Note that $f(\omega,.)$ is conditionally positive definite by Lemma 1.7. Putting now $\psi(\omega,g) = f(\omega,g)$ we have proved the theorem.

§ 8. The Bose - Einstein field of quantum mechanics

In this section we shall apply some of the results of the preceding section to obtain the socalled Bose-Einstein fields. In Theorem 7.2, we take G and X to coincide with a Hilbert space H, so that H acts on H by translation. Let A be a bounded operator in H. We define

$$L(x,y) = \langle Ax,Ay\rangle, \quad \text{for all } x,y \in H.$$

Then for any $z \in H$,

$$L(x+z,y+z) = L(x,y) + \{\langle Ax,Az\rangle + \tfrac{1}{2}\langle Az,Az\rangle\} + \{\langle Az,Ay\rangle + \tfrac{1}{2}\langle Az,Az\rangle\}.$$

Thus L is a conditionally positive definite kernel which is affine invariant. In fact L is positive definite. The function β of Theorem 7.2 is given by

$$\beta(z,x) = \langle Ax,Az\rangle + \tfrac{1}{2}\langle Az,Az\rangle, \text{ for all } z,x \in H.$$

The map v defined by $v(x) = Ax$ for all $x \in H$ is a cocycle of first order with origin 0 for the trivial representation of H acting in H. Let $\mathcal{H}$ be the subspace spanned by all the vectors $\{\text{Exp } Ax, x \in H\}$ in the Hilbert space Exp H. Let $z \to U_z$ be the projective representation of Theorem 7.2 defined by

$$U_z \text{Exp } Ax = \exp - \{\tfrac{1}{2}\langle Az,Az\rangle + \langle Ax,Az\rangle\}.\text{Exp } A(x+z). \tag{8.1}$$

It follows from Remark 6.3) that

$$U_{z_1}U_{z_2} = \{\exp i \operatorname{Im} \langle Az_1, Az_2\rangle\}.U_{z_1+z_2}, \text{ for all } z_1, z_2 \varepsilon H. \tag{8.2}$$

Since L is continuous it follows that the map $z \to U_z$ is a continuous projective representation with multiplier $\sigma(z_1,z_2) = \exp i \operatorname{Im} \langle Az_1, Az_2\rangle$. If A is a unitary operator, then (8.2) reduces to

$$U_{z_1}U_{z_2} = \{\exp i \operatorname{Im} \langle z_1, z_2\rangle\}.U_{z_1+z_2}, \text{ for all } z_1, z_2 \varepsilon H. \tag{8.3}$$

Further $\mathcal{H}$ is spanned by all Exp z, $z \varepsilon H$. Hence $\mathcal{H}$ = Exp H. Thus we have proved the following theorem.

<u>Theorem 8.1.</u> Let H be a Hilbert space and A be a unitary operator in H. Then there exists a projective unitary representation $z \to U_z$ of the additive group H in the Hilbertspace Exp H such that

$$U_z \operatorname{Exp} x = \exp\{-\tfrac{1}{2}||z||^2 - \langle x, Az\rangle\}.\operatorname{Exp}(x+Az) \quad \text{for all } x, z \varepsilon H,$$

$$U_{z_1}U_{z_2} = \{\exp i \operatorname{Im} \langle z_1, z_2\rangle\}.U_{z_1+z_2} \quad \text{for all } z_1, z_2 \varepsilon H.$$

Further the map $z \to U_z$ is continuous in H.

<u>Remark 8.3.</u> The relations (8.1) and (8.2) are the socalled Weyl commutation relations. The existence of such operators in the symmetric Fock spce is a well known result in quantum field theory. Any solution of (8.2) is called a Bose-Einstein field.

Remark 8.4. If $(\Omega,\mathcal{S})$ is a standard borel space and $P(E), E\epsilon\mathcal{S}$ is any projection valued measure operating in H, then the kernels

$$L(E,x,y) = \langle P(E)Ax,Ay\rangle$$

are conditionally positive definite for every fixed E and constitute an affine invariant additive family. Hence Theorem 7.7 is applicable. Thus the Bose-Einstein field constructed above is part of a factorisable representation of $S(\Omega,H)$.

Part II: Limit theorems for uniformly infinitesimal families of positive definite kernels

§9. Uniformly infinitesimal families of positive definite kernels

Throughout this section we shall assume that X is an arbitrary, but fixed set.

Definition 9.1. A positive definite kernel K on X×X is said to be normalised if $K(x,x) = 1$ for all $x\epsilon X$. A triangular array $\{K_{nj}, 1\leq j\leq k_n\}$ of normalised positive definite kernels on X×X is said to be uniformly infinitesimal if, for every $x,y\epsilon X$,

$$\lim_{n} \sup_{j} \{1 - |K_{nj}(x,y)|\} = 0. \tag{9.1}$$

Lemma 9.2. Let K be a normalised positive definite kernel on X×X, and let x_o be any fixed point. Then

$$|K(x,y) - K(x,x_o)K(x_o,y)| \leq 2 - |K(x,x_o)| - |K(x_o,y)|$$

for all $x,y \in X$.

Proof: By Theorem 1.2, there exists a map v from X into the unit sphere of a Hilbert space H such that $K(x,y) = \langle v(x),v(y)\rangle$ for all x,y. We select an orthonormal basis $\xi_o,\xi_1,\ldots$ of H such that $\xi_o = v(x_o)$. Since $|K(x,y)| \leq 1$ for all $x,y \in X$, we have

$$|K(x,y) - K(x,x_o)K(x_o,y)| = |\sum_{i\neq 0} \langle v(x),\xi_i\rangle\langle\xi_i,v(y)\rangle|$$

$$\leq \{\sum_{i\neq 0} |\langle v(x),\xi_i\rangle|^2\}^{\frac{1}{2}} . \{\sum_{i\neq 0} |\langle v(y),\xi_i\rangle|^2\}^{\frac{1}{2}}$$

$$\leq \{1 - |\langle v(x),v(x_o)\rangle|^2\}^{\frac{1}{2}} . \{1 - |\langle v(y),v(x_o)\rangle|^2\}^{\frac{1}{2}}$$

$$\leq 2 - |K(x,x_o)| - |K(x_o,y)|.$$

This proves the lemma.

Lemma 9.3. Let (K_n) be a sequence of positive definite kernels on X×X. Suppose $(\mathrm{Re}\, K_n)$ converges pointwise to a positive definite kernel k. Then the sequence (K_n) possesses a subnet which converges pointwise to a positive definite kernel K.

Proof: Since $\mathrm{Re}\, K_n$ converges at every point of X×X, it follows from Corollary 1.4 that $C(x,y) = \sup_n |K_n(x,y)| < \infty$ for every $x,y \in X$. Let, for every x,y, I(x,y) denote the set of complex numbers $\{z: |z| \leq C(x,y)\}$. The cartesian product $\mathcal{K} = \prod_{x,y\in X} I(x,y)$ is compact in the product topology, and every K_n can be identified with a point in $\mathcal{K}$ in a unique and obvious manner. Furthermore the product topology in $\mathcal{K}$ corresponds to the topology of pointwise convergence of the kernels K_n. Together with the compactness of

$\mathcal{K}$ this implies that the sequence (K_n) has a convergent subnet. Clearly, the limit K of such a subnet is a positive definite kernel on X×X. The lemma is proved.

Remark 9.4. If X is a topological space and if the kernel k is continuous, then K is continuous by Corollary 1.4. If, moreover, X is a separable metric space, then it is not difficult to see that there exists a subsequence of (K_n) converging to a continuous positive definite kernel K.

For the present we shall consider a fixed family $\{K_{nj},\ 1 \le j \le k_n\}$ of uniformly infinitesimal normalised positive definite kernels on X×X. We choose and fix a point x_o in X and write

$$\tilde{K}_{nj}(x,y) = K_{nj}(x,y)K_{nj}(x,x_o)^{-1}K_{nj}(x_o,y)^{-1} \tag{9.2}$$

$$K_n(x,y) = \prod_{j=1}^{k_n} K_{nj}(x,y) \tag{9.3}$$

$$\tilde{K}_n(x,y) = \prod_{j=1}^{k_n} \tilde{K}_{nj}(x,y) \tag{9.4}$$

$$X_o = \{x:\ \lim_n \inf |K_n(x,x_o)| > 0\}. \tag{9.5}$$

The kernels $\tilde{K}_{nj}$ and $\tilde{K}_n$ in (9.2) and (9.4) are only defined at those points where the denominator is nonzero.

Lemma 9.5. Let X_o be defined by (9.5). Then, for all $x,y \in X_o$, we have

$$\lim_n \inf |K_n(x,y)| > 0.$$

Proof: It is clear that $x \in X_o$ if and only if

$$\lim_n \sup \sum_{j=1}^{k_n} \{1 - |K_{nj}(x,x_o)|^2\} < \infty. \tag{9.6}$$

By Lemma 1.6, $|K_{nj}|^2$ is a normalised positive definite kernel for every n,j. Applying Lemma 9.2 to this kernel, we obtain

$$||K_{nj}(x,y)|^2 - |K_{nj}(x,x_o)|^2 |K_{nj}(x_o,y)|^2|$$

$$\leq 2 - |K_{nj}(x,x_o)|^2 - |K_{nj}(x_o,y)|^2.$$

Hence

$$\sum_j \{1-|K_{nj}(x,y)|^2\} \leq 4 \left(\sum_j \{1-|K_{nj}(x,x_o)|^2\} + \sum_j \{1-|K_{nj}(x_o,y)|^2\}\right). \tag{9.7}$$

Equations (9.6) and (9.7) imply that, for any $x,y \in X_o$,

$$\lim_n \sup \sum_j \{1 - |K_{nj}(x,y)|^2\} < \infty.$$

This completes the proof of the lemma.

Lemma 9.6. Let $\tilde{K}_{nj}$ and X_o be defined by (9.2) and (9.5), respectively. Then we have for all $x,y \in X_o$

$$\lim_n \sup \sum_j |1 - \tilde{K}_{nj}(x,y)| < \infty.$$

Proof: From the uniform infinitesimality of the K_{nj} we have for any two fixed $x,y \in X_o$,

$$|K_{nj}(x,x_o)| > \tfrac{1}{2}, \ |K_{nj}(x_o,y)| > \tfrac{1}{2}$$

for all j and all sufficiently large n. From Lemma 9.2 we have

$$|K_{nj}(x,y) - K_{nj}(x,x_o)K_{nj}(x_o,y)|$$

$$\leq 2 - |K_{nj}(x,x_o)| - |K_{nj}(x_o,y)|.$$

Thus, by (9.2),

$$|\tilde{K}_{nj}(x,y) - 1| \leq 4\ \{2 - |K_{nj}(x,x_o)| - |K_{nj}(x_o,y)|\}.$$

Since $|K_{nj}(x,y)| \leq 1$, an application of (9.5) completes the proof of the lemma.

<u>Lemma 9.7</u>. For all $x,y\varepsilon X_o$,

$$\lim_n \sup \sum_j |\mathrm{Log}\ \tilde{K}_{nj}(x,y) + 1 - \tilde{K}_{nj}(x,y)| = 0, \tag{9.8}$$

where Log stands for the principal value of the logarithm.

<u>Proof</u>: We have

$$|\mathrm{Log}\ x + 1 - x| \leq |1 - x|^2 \quad \text{for } |1 - x| < \tfrac{1}{2}.$$

Hence the left hand side of (9.8) is less than or equal to

$$\lim_n \sup \sum_j |1 - \tilde{K}_{nj}(x,y)|^2$$

$$\leq \lim_n \sup \{\sup_j |1 - \tilde{K}_{nj}(x,y)|\}.\sum_j |1 - \tilde{K}_{nj}(x,y)|.$$

The uniform infinitesimality of $\{K_{nj}\}$ and Lemma 9.6 imply Lemma 9.7. This completes the proof.

<u>Theorem 9.8</u>. Let X be a set and let $\{K_{nj}, 1\leq j\leq k_n\}$ be a uniformly infinitesimal family of normalised positive definite kernels on X×X. Let $x_o\varepsilon X$ be an arbitrary but fixed point and let X_o be defined as in (9.5). Then for any $x,y\varepsilon X_o$, we have

$$\lim_n \sup |\exp \sum_j \{\tilde{K}_{nj}(x,y) - 1\} - \prod_j \tilde{K}_{nj}(x,y)| = 0,$$

where $\tilde{K}_{nj}$ is given by (9.2).

<u>Proof</u>: This is an immediate consequence of Lemma 9.7.

Remark 9.9. Suppose we write

$$L_{nj}(x,y) = \tilde{K}_{nj}(x,y) - 1 \tag{9.9}$$

$$L_n(x,y) = \prod_{j=1}^{k_n} L_{nj}(x,y) \tag{9.10}$$

for all x,y, for which the above expressions are defined. For every $x,y\varepsilon X$, $\tilde{K}_{nj}(x,y)$, $L_{nj}(x,y)$ and $L_n(x,y)$ exist for all sufficiently large n and all j. Furthermore, each $\tilde{K}_{nj}$ and hence each $\tilde{K}_n$ is positive definite in its domain. Consequently each L_{nj} and each L_n is conditionally positive definite in its domain. Since

$$L_{nj}(x,y) = L_{nj}(x,y) - L_{nj}(x,x_o) - L_{nj}(x_o,y) + L_{nj}(x_o,x_o)$$

wherever it is defined, it follows from Lemma 1.7 that each L_{nj} and hence each L_n is positive definite in its domain. We can call exp L_n the accompanying kernel for the sequence $\{K_n\}$ defined by (9.3). In analogy with Gnedenko and Kolmogorov, 'Limit distributions of sums of independent random variables', § 24, Chapter 4, we may call Theorem 9.7 the accompanying kernel theorem. We shall see later that this will yield in particular the central limit theorems of probability theory.

Theorem 9.10. Let X be a set and let $\{K_{nj}, 1\leq j\leq k_n\}$ be a uniformly infinitesimal family of normalised positive definite kernels on $X\times X$. Suppose

$$\lim_n \prod_{j=1}^{k_n} K_{nj}(x,y) = K(x,y) \tag{9.11}$$

exists for all $x,y\varepsilon X$. Let $x_o\varepsilon X$ be an arbitrary, but fixed point, and let X_o be defined as $X_o = \{x: K(x,x_o) \neq 0\}$. Then there exists a positive definite kernel M on $X_o \times X_o$ such that

$$K(x,y) = K(x,x_o)K(x_o,y).\exp M(x,y)$$

for all $x,y \in X_o$.

Proof: We define L_{nj} and L_n by (9.9) and (9.10). Then (9.11) implies that

$$\lim_n \prod_j \tilde{K}_{nj}(x,y) = K(x,y)K(x,x_o)^{-1}K(x_o,y)^{-1}$$

for all $x,y \in X_o$. From Theorem 9.8 we have for any $x,y \in X_o$,

$$K(x,y)K(x,x_o)^{-1}K(x_o,y)^{-1} = \lim_n \exp L_n(x,y).$$

From the convergence of $\prod_j |K_{nj}(x,y)|$ we conclude from the same theorem that Re L_n converges pointwise on $X_o \times X_o$. For each n, we denote by D_n the domain of L_n, and we put $S_n = \bigcap_{n \leq m} D_m$. Clearly, $\bigcup S_n \supset X_o \times X_o$. It follows from Remark 9.9 that L_m is positive definite on S_n for $m \geq n$. Hence by Lemma 9.3, the sequence (L_m) possesses a subnet converging on $S_n \cap X_o \times S_n \cap X_o$. Since $X_o \times X_o$ is the union of the sets $S_n \cap X_o \times S_n \cap X_o$, we can choose a subnet of (L_n) which converges on $X_o \times X_o$. Let M be the limit of this subnet. Clearly M fulfills our reqirements. The theorem is proved.

Remark 9.11. If we assume X to be a topological space and the kernels K_{nj} and K to be continuous, then the continuity of $\log |K(x,y)K(x,x_o)^{-1}K(x_o,y)^{-1}| = \text{Re } M(x,y)$ implies the continuity of M by Corollary 1.4.

§10. Uniformly infinitesimal families of projectively invariant positive definite kernels

Throughout this section we shall assume that X is a fixed set and G is a group acting on X.

Lemma 10.1. Let K be a normalised projectively invariant positive definite kernel on X×X. Then $K(gx,gy) = \alpha(g,x)\overline{\alpha(g,y)}\, K(x,y)$, where $|\alpha(g,x)| = 1$ for all $g\epsilon G$, $x\epsilon X$. If $K(x,y) \neq 0$ for some x,y, then $K(gx,gy) \neq 0$ for all $g\epsilon G$.

Proof: This follows immediately from Definition 2.1 and 9.1.

For the following we choose and fix a uniformly infinitesimal family $\{K_{nj}, 1 \leq j \leq k_n\}$ of continuous projectively invariant positive definite normalised kernels on X×X. We choose and fix furthermore a point x_o in X and define $\tilde{K}_{nj}$, K_n, $\tilde{K}_n$ and X_o exactly as in (9.2) - (9.5).

Lemma 10.2. The set $G_o = \{g: gx_o \epsilon X_o\}$ is a subgroup of G.

Proof: Suppose g_1 and g_2 are elements of G_o. Then, by Lemma 9.5,

$$\liminf_n |K_n(g_1x_o,g_2x_o)| > 0.$$

By Lemma 10.1, $|K_n(g_1x_o,g_2x_o)| = |K_n(g_2^{-1}g_1x_o,x_o)|$. This shows that $g_2^{-1}g_1 \epsilon G_o$ and completes the proof of the lemma.

Lemma 10.3. Let G_o be defined as in Lemma 10.2. Then for any $x\epsilon X_o, g\epsilon G_o$, we have $gx\epsilon X_o$.

Proof: Since $x \in X_o$, we have

$$\lim_n \inf |K_n(gx,gx_o)| = \lim_n \inf |K_n(x,x_o)| > 0.$$

By the definition of G_o,

$$\lim_n \inf |K_n(gx_o,x_o)| > 0, \text{ for } g \in G_o.$$

Hence, by Lemma 9.5,

$$\lim_n \inf |K_n(gx,x_o)| > 0, \text{ for } g \in G_o.$$

This completes the proof of the lemma.

Theorem 10.4. Let G be a group acting on a set X, and let $\{K_{nj}, 1 \leq j \leq k_n\}$ be a uniformly infinitesimal family of normalised projectively invariant positive definite kernels on X×X. Suppose

$$\lim_n \prod_j K_{nj}(x,y) = K(x,y)$$

exists for all $x,y \in X$. Let x_o be a fixed point in X, $G_o = \{g: K(gx_o,x_o) \neq 0\}$, and $X_o = \{x: K(x,x_o) \neq 0\}$. Then G_o is a subgroup of G which leaves X_o invariant. Further there exists a G_o-affine invariant positive definite kernel M on $X_o \times X_o$ such that

$$K(x,y) = K(x,x_o)K(x_o,y).\exp M(x,y) \tag{10.1}$$

for all $x,y \in X_o$.

Proof: That G_o is a subgroup follows from Lemma 10.2, and the invariance of X_o under G_o follows from Lemma 10.3. If we define $\tilde{K}_{nj}$ by (9.2) and L_n by (9.10) we see from the proof of Theorem 9.10 that (10.1) holds with M as a limit of a subnet of (L_n) on

$X_o \times X_o$. We shall show that M must be affine invariant under G_o. For any $x,y \varepsilon X_o$, $g \varepsilon G_o$, we have for sufficiently large n and all j,

$$\tilde{K}_{nj}(gx,gy) = \alpha_{nj}(g,x)\overline{\alpha_{nj}(g,y)}\tilde{K}_{nj}(x,y) \qquad (10.2)$$

for some $\alpha_{nj}(g,x)$. By Lemma 9.6 and uniform infinitesimality, we have

$$\lim_n \sup \sum_j |\tilde{K}_{nj}(gx,gy) - 1| < \infty,$$

$$\lim_n \sup \sum_j |\tilde{K}_{nj}(x,y)^{-1} - 1| < \infty,$$

for every $x,y \varepsilon X_o$, $g \varepsilon G_o$. Hence

$$\lim_n \sup \sum_j |\alpha_{nj}(g,x)\overline{\alpha_{nj}(g,y)} - 1| < \infty. \qquad (10.3)$$

(10.2), (10.3) and uniform infinitesimality together imply that

$$\lim_n \sup \sum_j |\tilde{K}_{nj}(gx,gy) - \tilde{K}_{nj}(x,y) - \alpha_{nj}(g,x)\overline{\alpha_{nj}(g,y)} + 1| = 0.$$

From the last equation and from the convergence of a suitable subnet of $(\sum_j \{\tilde{K}_{nj} - 1\})$ to M we conclude the convergence of a suitably chosen subnet of $(\sum_j \{\alpha_{nj}(g,x)\overline{\alpha_{nj}(g,y)} - 1\})$. Let us denote the limit of this subnet by $P(g,x,y)$. Now (10.2), (10.3) and uniform infinitesimality imply that

$$P(g,x,y) = P(g,x,x_o) + P(g,x_o,y) - P(g,x_o,x_o)$$

for all $x,y \varepsilon X_o$, $g \varepsilon G_o$. Putting now $\beta(g,x) = P(g,x,x_o) - \frac{1}{2} P(g,x_o,x_o)$, we see that for all $x,y \varepsilon X_o, g \varepsilon G_o$,

$$M(gx,gy) = M(x,y) + \beta(g,x) + \overline{\beta(g,y)}.$$

This shows the affine invariance of M under the action of G_o.

Remark 10.5. Combining the Theorems 10.4 and 3.4, we see that the limiting kernel K of Theorem 10.4 satisfies the identity

$$K(x,y) = K(x,x_o)K(x_o,y).\exp \langle v(x),v(y)\rangle$$

for all x,y such that $K(x,x_o) \neq 0$, $K(x_o,y) \neq 0$, where v is a first order cocycle with origin x_o for some unitary representation V of $G_o = \{g: K(gx_o,x_o) \neq 0\}$.

Remark 10.6. If we assume that X is a topological space and G a topological group acting continuously on X, and if we further assume that the kernel K is continuous, then we conclude from Remark 9.4, that M is a continuous affine invariant positive definite kernel on the open subset $X_o \times X_o$ of $X \times X$, and G_o is an open subgroup of G.

Remark 10.7. If all the kernels K_{nj} in Theorem 10.4 are invariant we can write $\phi_{nj}(g) = K_{nj}(gx_o,x_o)$ and $\phi(g) = K(gx_o,x_o)$. Clearly every ϕ_{nj} and ϕ is a positive definite function on G, which is normalised. The above equations imply that for every $g_1,g_2 \in G_o$,

$$\phi(g_1g_2)\phi(g_1)^{-1}\phi(g_2)^{-1} = \exp \langle\delta(g_2),\delta(g_1^{-1})\rangle,$$

where $\delta(g) = v(gx_o)$, $g \in G_o$. Note that for every $g,h \in G_o$,

$$V_g\delta(h) = \delta(gh) - \delta(g),$$

where v and V have the same meaning as in Remark 10.5. Outside G_o, $\phi \equiv 0$. By looking at the proof of Theorem 10.4, it is not difficult to see that $\langle\delta(g_2),\delta(g_1^{-1})\rangle$ is the limit of a suitable subnet of the sequence ($\sum_j \{\phi_{nj}(gh)\phi_{nj}(g)^{-1}\phi_{nj}(h)^{-1} - 1\}$). The question arises whether the above limit is of the form $\psi(gh) - \psi(g) - \psi(h)$ for some conditionally positive definite function ψ on G_o We shall answer this question in the next section.

§11. Uniformly infinitesimal families of normalised positive definite functions

Let G be a group. We shall now analyse the properties of limits of products of positive definite functions on G which constitute a uniformly infinitesimal family.

Definition 11.1. A triangular array of normalised positive definite functions $\{\phi_{nj}, 1 \leq j \leq k_n\}$ on G is called uniformly infinitesimal if the family of kernels $\{K_{nj}, 1 \leq j \leq k_n\}$, $K_{nj}(g,h) = \phi_{nj}(h^{-1}g)$, is uniformly infinitesimal.

Theorem 11.2. Let G be a group and $\{\phi_{nj}, 1 \leq j \leq k_n\}$ a uniformly infinitesimal family of normalised positive definite functions on G. Suppose the sequence $\phi_n = \prod_j \phi_{nj}$ converges pointwise to a normalised positive definite function ϕ on G. Then the set $G_o = \{g: \phi(g) \neq 0\}$ is a subgroup of G. Further there exists a unitary representation V of G_o in a Hilbert space H and a map δ from G_o into H such that

$$V_g \delta(h) = \delta(gh) - \delta(g) \text{ for all } g,h \in G_o, \qquad (11.1)$$

$$\phi(gh)\phi(g)^{-1}\phi(h)^{-1} = \exp \langle \delta(h), \delta(g^{-1}) \rangle \text{ for all } g,h \in G_o. \qquad (11.2)$$

If X is a topological space and if the functions ϕ_{nj} and ϕ are continuous, then the subgroup G_o is open, the representation V of G_o is weakly continuous, and the map δ is continuous.

Proof: If we consider G to be acting on itself by left translation, then the kernels $K_{nj}(g,h) = \phi_{nj}(h^{-1}g)$ are invariant under G. An application of Theorem 10.4 and the Remarks 10.6 and 10.7 proves the theorem.

In classical probability theory the following problem plays an important role: What is the form of all continuous positive definite functions on the real line, which arise as limits of uniformly infinitesimal families of continuous positive definite functions in the way described in Theorem 11.2? The description of these functions is given in two steps: First it is shown that any such function is the exponential of a continuous conditionally positive definite function, and then a formula for all conditionally positive definite functions (the Levy-Khinchine formula) is given. Our next aim is to prove the first of these two assertions on any group. We shall show that any normalised positive definite function ϕ which arises in the above manner is of the form $\phi = \chi.\exp \psi$ on G_o, where G_o is the subgroup where ϕ is nonzero, χ a homomorphism from G_o into the group of complex numbers of modulus unity, and ψ a normalised conditionally positive definite function on G_o. In order to do this we first need a few lemmas on second order cocycles.

Lemma 11.3. Let A be an abelian group, A_o a subgroup, and θ_o a homomorphism from A_o into the additive group of complex numbers.

Then θ_o can be extended to a homomorphism defined on the whole of A. If A is a topological group, A_o is open, and θ_o is continuous on A_o, then θ is continuous.

Proof: An application of Zorn's Lemma shows that it is enough to extend a homomorphism defined on A_o to a homomorphism on the group generated by A_o and an element $g_o \notin A_o$. If there exists an integer m such that $g_o^m \in A_o$, and if m_o is the smallest positive integer of this kind, then we put

$$\theta(g_o^n g) = \theta_o(g) + \frac{n}{m_o}\theta_o(g_o^{m_o})$$

for any n and any $g \in A_o$. If no such m exists, we define

$$\theta(g_o^n g) = \theta_o(g).$$

A straightforward verification shows that θ is a homomorphism on the bigger group. If A is a topological group, A_o is open, and θ_o continuous, then θ is obviously also continuous. This proves the lemma.

Lemma 11.4. Let A be an abelian group and let s be a second order cocycle on A×A with values in the additive group of complex numbers. Assume that s is symmetric in its arguments. Then there exists a complex valued function α on A such that

$$s(g,h) = \alpha(gh) - \alpha(g) - \alpha(h) \qquad (11.3)$$

for all $g,h \in A$.

Proof: We define a central extension of the complex numbers C by A in the usual way as the set A×C with the group operation $(g,x)\cdot(g',x') = (gg',x+x'+s(g,g'))$ for any $g,g' \in A$, $x,x' \in C$.

Since $s(g,g') = s(g',g)$ for all g,g', this extension - we shall denote it by A_s - is again an abelian group. Consider the subgroup $C_e = \{(e,x),\ x\in C\}$, where e denotes the identity in A. The map θ_o: $(e,x)\to x$ is a complex valued homomorphism on C_e. By Lemma 11.3 it can be extended to a homomorphism θ on A_s. Define now $\alpha(g) = -\theta(g,0)$ for any $g\in A$. Obviously we have $\alpha(gh) - \alpha(g) - \alpha(h) = s(g,h)$ for all $g,h\in A$. This proves the lemma.

Lemma 11.5. Let A be a locally compact second countable abelian group and let s be a borel measurable second order cocycle on A×A with values in the additive group of complex numbers. If s is symmetric in its arguments, then the function α fulfilling (11.3) can be chosen as borel measurable.

Proof: We define A_s as in the proof of Lemma 11.4, and furnish it with the Weil topology in the usual way (for details see the Lecture Note volume 'Multipliers on locally compact groups' by the first named author). A_s is then a locally compact second countable abelian group. Let A_s^o be an open subgroup of A_s generated by some relatively compact neighbourhood of the identity (e,0) in A_s. By structure theory, A_s^o is of the form $K\times Z^m\times R^n$, where K is a compact group, and Z^m and R^n are the cartesian products of m copies of the integers and n copies of the real line, respectively. Let again C_e be the subgroup $\{(e,x),\ x\in C\}$. It is easy to see that C_e forms a two dimensional subspace of the R^n part of A_s^o. Hence there exists a continuous homomorphism θ_o from A_s^o into C such that $\theta_o(e,x) = x$ for every $x\in C$. By Lemma 11.3 we can extend θ_o to a continuous homomorphism θ on A_s.

Since the map $g \to (g,0)$ from A into A_s is measurable, we can define a measurable function α by $\alpha(g) = -\theta(g,0)$ for every $g \in A$. Obviously α fulfils (11.3). This proves the lemma.

Lemma 11.6. Let A and B be locally compact second countable groups, B abelian, and let s be a continuous second order cocycle on $A \times A$ with values in B. Assume there exists a borel map α from A into B such that $\alpha(gh)\alpha(g)^{-1}\alpha(h)^{-1} = s(g,h)$ for all $g,h \in A$. Then α is continuous.

Proof: Let A_s be the set $\{(a,b),\ a \in A,\ b \in B\}$ with the group operation $(a,b).(a',b') = (aa',bb's(a,a'))$ for $a,a' \in A$, $b,b' \in B$, furnished with the product topology. A_s is then a locally compact second countable group. We define a map β from the product group $A \times B$ into A_s by

$$\beta(a,b) = (a,b\alpha(a)),\ a \in A,\ b \in B.$$

β is measurable. Since β is a group homomorphism of locally compact second countable groups we conclude that β is continuous. Hence α is continuous. The lemma is proved.

Lemma 11.7. Let G be a compact group and s a continuous second order cocycle on $G \times G$ with values in the additive group of complex numbers. Then there exists a continuous complex valued function α on G such that

$$\alpha(gh) - \alpha(g) - \alpha(h) = s(g,h)$$

for all $g,h \in G$.

Proof: Let dg denote integration with respect to the normalised Haar measure on G. Then we have from the cocycle identity

$$\int s(g_1,g)\,dg + \int s(g_2,g)\,dg = \int s(g_1g_2,g)\,dg + s(g_1,g_2)$$

Putting $\alpha(g) = -\int s(g,h)\,dh$ for every $g\varepsilon G$, we have proved the lemma.

The commutator subgroup of a group G is defined as the subgroup generated by all elements $ghg^{-1}h^{-1}$, $g,h\varepsilon G$. We denote by D_o the commutator subgroup of the group $G_o \subset G$. D_o is a normal subgroup of G_o and the quotient G_o/D_o is abelian. With these notations we have:

Lemma 11.8. For every $g\varepsilon D_o$, the following two relations hold:

$$\lim_n \sup \sum_j |\phi_{nj}(g)-1| < \infty,$$
$$\lim_n \sup_j |\phi_{nj}(g)-1| = 0. \qquad (11.4)$$

Proof: First we show that the set of all elements, for which the first relation in (11.4) holds, is a subgroup of G_o. We have an inequality

$$|\phi_{nj}(gh) - 1| \leq |\phi_{nj}(gh)\phi_{nj}(g)^{-1}\phi_{nj}(h)^{-1} - 1| + |\phi_{nj}(g) - 1| + |\phi_{nj}(h) - 1|.$$

Hence

$$\lim_n \sup \sum_j |\phi_{nj}(gh) - 1|$$

$$\leq \lim_n\sup \sum_j |\phi_{nj}(gh)\phi_{nj}(g)^{-1}\phi_{nj}(h)^{-1} - 1|$$

$$+ \lim_n\sup \sum_j |\phi_{nj}(g) - 1| + \lim_n\sup \sum_j |\phi_{nj}(h) - 1|.$$

Our assertion follows now from Lemma 9.6, where we put

$$\widetilde{K}_{nj}(h,g^{-1}) = \phi_{nj}(gh)\phi_{nj}(g)^{-1}\phi_{nj}(h)^{-1}.$$

Therefore it will be sufficient to show that the first inequality in (11.7) holds for all $ghg^{-1}h^{-1}$, $g,h\in G_o$. Applying the inequality

$$\begin{aligned}|\phi_{nj}(ghg^{-1}h^{-1}) - 1| \leq\ & |\phi_{nj}(ghg^{-1}h^{-1})\phi_{nj}(gh)^{-1}\phi_{nj}(g^{-1}h^{-1}) - 1| \\ &+ |\phi_{nj}(gh)\phi_{nj}(g)^{-1}\phi_{nj}(h)^{-1} - 1| \\ &+ |\phi_{nj}(g^{-1}h^{-1})\phi_{nj}(g^{-1})^{-1}\phi_{nj}(h^{-1})^{-1} - 1| \\ &+ ||\phi_{nj}(g)|^2 - 1| + ||\phi_{nj}(h)|^2 - 1|,\end{aligned}$$

we get $\lim_n\sup \sum_j |\phi_{nj}(ghg^{-1}h^{-1}) - 1| < \infty$ by Lemma 9.6 and (9.6). The second relation follows in a similar manner from the two inequalities above, Lemma 9.2, and uniform infinitesimality. The lemma is proved.

Let us now turn back to our problem. We are given a uniformly infinitesimal family $\{\phi_{nj}, 1\leq j\leq k_n\}$ of normalised positive definite functions on a group G, and we assume that $\phi_n = \prod_j \phi_{nj}$ converges pointwise to a positive definite function ϕ on G. Let G_o be the subgroup $\{g: \phi(g) \neq 0\}$ and δ the cocycle defined by (11.1) and (11.2) taking values in some Hilbert space H. We put

$$N(g,h) = \langle\delta(h),\delta(g^{-1})\rangle, \quad g,h\varepsilon G_o. \tag{11.5}$$

Remark 10.7 implies that we can choose a subnet, denoted by $(\phi_\alpha = \prod_j \phi_{\alpha j})$, of the sequence (ϕ_n) such that

$$N(g,h) = \lim_\alpha \sum_j \{\phi_{\alpha j}(gh)\phi_{\alpha j}(g)^{-1}\phi_{\alpha j}(h)^{-1} - 1\} \tag{11.6}$$

for all $g,h\varepsilon G_o$. We first note that (11.1) and (11.5) imply by straighforward computation, that N is a second order cocycle on $G_o \times G_o$. Indeed we have

$$\left.\begin{aligned} N(g_1,g_2) + N(g_1g_2,g_3) &= N(g_1,g_2g_3) + N(g_2,g_3) \\ &\quad \text{for all } g_1,g_2,g_3\varepsilon G_o \\ N(g,e) = N(e,g) &= 0 \text{ for all } g\varepsilon G_o. \end{aligned}\right\} \tag{11.7}$$

From (11.4) we conclude that for every $g\varepsilon D_o$,

$$C(g) = \sup_n \sum_j |\phi_{nj}(g) - 1| < \infty.$$

Let I(g) denote the set of complex numbers $\{z: |z| \leq C(g)\}$. The cartesian product $\mathcal{H} = \prod_{g\varepsilon D_o} I(g)$ is compact in the product topology, and we may identify the restriction of each function $\sum_j \{\phi_{nj} - 1\}$ to D_o with a point in $\mathcal{H}$, just as in the proof of Lemma 9.3. Since the topology on $\mathcal{H}$ corresponds to the topology of pointwise convergence of the functions $\sum_j \{\phi_{nj} - 1\}$ on D_o, we conclude from the compactness of $\mathcal{H}$ that there exists a subnet $(\phi_\beta = \prod_j \phi_{\beta j})$ of the net (ϕ_α) such that $\sum_j \{\phi_{\beta j} - 1\}$ converges on D_o to a limit f. Clearly, f is a normalised conditionally positive definite function on D_o. With these notations we have a few more lemmas.

Lemma 11.9. For any $g_1,g_2\varepsilon D_o$, we have

$$f(g_1g_2) - f(g_1) - f(g_2) = N(g_1,g_2).$$

Proof: Consider the relation

$$\lim_n\sup \sum_j |\phi_{nj}(g_1g_2)\phi_{nj}(g_1)^{-1}\phi_{nj}(g_2)^{-1} - 1 - \{\phi_{nj}(g_1g_2) - \\ - \phi_{nj}(g_1) - \phi_{nj}(g_2) + 1\}|$$

$$\leq \lim_n\sup \sum_j \{|\phi_{nj}(g_1)\phi_{nj}(g_2) - 1|.|\phi_{nj}(g_1g_2)\phi_{nj}(g_1)^{-1}\phi_{nj}(g_2)^{-1} - 1| \\ + |\phi_{nj}(g_1) - 1|.|\phi_{nj}(g_2) - 1|\} = 0,$$

by Hölder's inequality, uniform infinitesimality, and the Lemmas 9.6 and 11.8, for any $g_1,g_2\varepsilon D_o$.

Furthermore we have, from the above relation,

$$|f(g_1g_2) - f(g_1) - f(g_2) - N(g_1,g_2)|$$

$$\leq \lim_n\sup \sum_j |\phi_{nj}(g_1g_2)\phi_{nj}(g_1)^{-1}\phi_{nj}(g_2)^{-1} - 1 - \phi_{nj}(g_1g_2) + 1 \\ + \phi_{nj}(g_1) - 1 + \phi_{nj}(g_2) - 1| = 0,$$

for any $g_1,g_2\varepsilon D_o$. This proves the lemma.

Lemma 11.10. Let $h\varepsilon D_o$. Then, for any $g\varepsilon G_o$, we have

$$f(g^{-1}hg) = f(h) + N(g^{-1}h,g) + N(g^{-1},h) - N(g^{-1},g) \qquad (11.8)$$

Proof: This follows from

$$\phi_{nj}(g^{-1}hg) - \phi_{nj}(h) -$$

$$- (\{\phi_{nj}(g^{-1}hg)\phi_{nj}(g^{-1}h)^{-1}\phi_{nj}(g)^{-1} - 1\}.$$

$$.\{\phi_{nj}(g^{-1}h)\phi_{nj}(g^{-1})^{-1}\phi_{nj}(h)^{-1}\}.|\phi_{nj}(g)|^2.\phi_{nj}(h)$$

$$+ \{\phi_{nj}(g^{-1}h)\phi_{nj}(g^{-1})^{-1}\phi_{nj}(h)^{-1} - 1\}.|\phi_{nj}(g)|^2.\phi_{nj}(h)$$

$$+ \{|\phi_{nj}(g)|^2 - 1\}.\phi_{nj}(h)\} = 0.$$

Summing over j and taking the limit with respect to (ϕ_β) together with (11.4) gives the result. The lemma is proved.

Lemma 11.11. f can be extended to a complex valued function ψ on G_o which satisfies

$$N(g_1,g_2) = \psi(g_1g_2) - \psi(g_1) - \psi(g_2) \tag{11.9}$$

for every $g_1,g_2 \varepsilon G_o$. If, moreover, G is a locally compact and second countable topological group, N is continuous on $G_o \times G_o$, and f is continuous on D_o, then there exists a continuous extension ψ of f to the whole of G_o.

Proof: Choose a subset F of G_o which intersects each coset of D_o in G_o in exactly one point, and which contains the identity element e. Every element g of G_o can then be written in a unique manner as a product $g = g^*.h$, where $g^* \varepsilon F$, $h \varepsilon D_o$. This allows us to define a function b on G_o by putting for any $g \varepsilon G_o$,

$$b(g) = f(h) + N(g^*,h),$$

where $g = g^*.h$, $g^* \varepsilon F$, $h \varepsilon D_o$, is the decomposition described above. Since F contains e, b is an extension of f.

Put now, for any $g_1,g_2 \varepsilon G_o$,

$$s(g_1,g_2)=b(g_1g_2) - b(g_1) - b(g_2).$$

Clearly, s is a second order cocycle on $G_o \times G_o$. $t = N - s$ is also a second order cocycle on $G_o \times G_o$. Using the results of the Lemmas 11.9 and 11.10, one can show by straightforward computation , that t is constant on each coset of $D_o \times D_o$ in $G_o \times G_o$, and furthermore, that t is symmetric in its arguments. We can thus define a second order cocycle $\tilde{t}$ on $G_o/D_o \times G_o/D_o$ by putting $\tilde{t}(\tilde{g}_1,\tilde{g}_2) = t(g_1,g_2)$ whenever $\tilde{g}_1$ and $\tilde{g}_2$ are the images of g_1 and g_2 under the natural homomorphism from G_o onto G_o/D_o. Since $\tilde{t}$ is symmetric in its arguments and since G_o/D_o is abelian, an appliccation of Lemma 11.4 yields the existence of a complex valued function $\tilde{d}$ on G_o/D_o with the property, that for any $\tilde{g}_1,\tilde{g}_2 \varepsilon G_o/D_o$,

$$\tilde{t}(\tilde{g}_1,\tilde{g}_2) = \tilde{d}(\tilde{g}_1\tilde{g}_2) - \tilde{d}(\tilde{g}_1) - \tilde{d}(\tilde{g}_2).$$

Clearly, $\tilde{d}(\tilde{e}) = 0$, where $\tilde{e}$ denotes the identity in G_o/D_o, since $\tilde{t}(\tilde{e},\tilde{e}) = 0$. We define a function d on G_o by putting $d(g) = \tilde{d}(\tilde{g})$ whenever $\tilde{g}$ is the image of g under the natural homomorphism from G_o onto G_o/D_o. Putting now $\psi = b + d$, we see that ψ extends f and fulfils (11.9). This proves the first part of the lemma.

Assume now that G is locally compact and second countable, and that the functions N and f are continuous on their domains. It is an easy consequence of the continuity of N that f can be extended to a continuous function on the closure $\bar{D}_o$ of D_o. If we denote the extended function by the same symbol f, we have for any $g,h \varepsilon \bar{D}_o$, $f(gh) - f(g) - f(h) = N(g,h)$ by Lemma

11.9 and continuity. It follows from a theorem of Kuratowski that there exists a one-one borel map ρ from the quotient $G_o/\bar{D}_o$ into G_o such that $\rho(\tilde{e}) = e$, where $\tilde{e}$ denotes the identity in $G_o/\bar{D}_o$, and $\pi\rho(\tilde{g}) = \tilde{g}$ for any $\tilde{g}\varepsilon G_o/\bar{D}_o$, where π denotes the natural homomorphism from G_o onto $G_o/\bar{D}_o$. For a proof of this the reader may refer to 'Probability measures on metric spaces' by the first author. We choose and fix such a map ρ. As before, we define a complex valued function b on G_o by putting, for every $g\varepsilon G_o$,

$$b(g) = f(\rho\pi(g)^{-1}g) + N(\rho\pi(g),\rho\pi(g)^{-1}g).$$

b is borel measurable, and $s(g_1,g_2) = b(g_1g_2) - b(g_1) - b(g_2)$ $g_1,g_2\varepsilon G_o$, is a borel measurable second order cocycle on $G_o\times G_o$. As before, $t = N - s$ is a second order cocycle on $G_o\times G_o$, which is constant on the cosets of $\bar{D}_o\times\bar{D}_o$ in $G_o\times G_o$, symmetric in its arguments, and in addition borel measurable. The rest of the proof is exactly as in the first part, using this time the Lemmas 11.6 and 11.7. The proof is complete.

Lemma 11.12. $f(g) = \lim_{\alpha} \sum_{j}\{\phi_{\alpha j}(g) - 1\}$ exists for every $g\varepsilon D_o$.
Proof: Assume there exist two different subnets (ϕ_β) and $(\phi_{\beta'})$ of (ϕ_α), such that $f = \lim_{\beta} \sum_{j} \{\phi_{\beta j} - 1\}$ and $\lim_{\beta'} \sum_{j} \{\phi_{\beta' j} - 1\} = f'$ both exist on D_o and that $f(g_o) \neq f'(g_o)$ for some $g_o\varepsilon D_o$. By Lemma 11.11, both f and f' can be extended to functions ψ and ψ' on G_o which satisfy (11.9) there. The difference $\psi-\psi'$ must therefore be a complex homomorphism of G_o. But every such homomorphism vanishes on D_o, which contradicts our assumption

that $f(g_o) \neq f'(g_o)$. Hence $f = f'$, and $\lim\limits_{\alpha} \sum\limits_{j} \{\phi_{\alpha j} - 1\}$ exists on D_o. The lemma is proved.

<u>Theorem 11.13</u>. Let G be a group and $\{\phi_{nj}, 1 \leq j \leq k_n\}$ a uniformly infinitesimal family of normalised positive definite functions on G. Suppose $\prod\limits_{j} \phi_{nj} = \phi_n$ converges pointwise to a positive definite function ϕ on G. Then there exists a subgroup G_o of G, a normalised conditionally positive definite function ψ on G_o, and a homomorphism χ from G_o into the multiplicative group of complex numbers of modulus 1, such that

$$\begin{aligned} \phi(g) &= \chi(g).\exp \psi(g) \text{ for every } g \in G_o, \\ &= 0 \text{ elsewhere.} \end{aligned} \tag{11.10}$$

<u>Proof</u>: By Lemma 11.12 and the discussion preceding Lemma 11.9 we can choose a subnet (ϕ_α) of the sequence (ϕ_n) such that $\lim\limits_{\alpha} \sum\limits_{j} \{\phi_{\alpha j}(gh)\phi_{\alpha j}(g)^{-1}\phi_{\alpha j}(h)^{-1} - 1\} = N(g,h)$ for every $g,h \in G_o$, where $N(g,h)$ fulfils (11.5) for some cocycle δ. Furthermore, by Lemma 11.12, $f = \lim\limits_{\alpha} \sum\limits_{j} \{\phi_{\alpha j} - 1\}$ exists on D_o, the commutator subgroup of G_o. Using Lemma 11.11 we extend f to a function ψ on G_o, which fulfils (11.9) for all $g_1, g_2 \in G_o$. Since $\psi(e) = 0$ and since $N(g,h) = \overline{N(h^{-1},g^{-1})}$, we have $\psi(g) = \overline{\psi(g^{-1})}$ for every $g \in G_o$. Lemma 1.7 implies now that ψ is conditionally positive definite, since $M(g,h) = N(h^{-1},g)$ is a positive definite kernel by Theorem 10.4. Since furthermore, by Remark 10.7 and by (11.9), $\phi(g_1g_2)\phi(g_1)^{-1}\phi(g_2)^{-1} = \exp\{\psi(g_1g_2) - \psi(g_1) - \psi(g_2)\}$ for all $g_1,g_2 \in G_o$, Re $\psi - \log|\phi|$ must be a homomorphism from G_o into the additive group of real numbers. Altering ψ by this

homomorphism if necessary (this does not affect the other properties of ψ) we may assume that Re $\psi = \log |\phi|$. But such a choice of ψ implies that exp ψ and ϕ differ only by a homomorphism χ of modulus unity. The theorem is proved.

Remark 11.14. It was claimed in [17] by the second author, that on a locally compact second countable group G and for continuous functions ϕ_{nk} and ϕ fulfilling the conditions of Theorem 11.13, one can always find continuous functions ψ and χ satisfying (11.10). However, the proof given in [17], that ψ and χ are continuous, contains an error. We shall now investigate the problem of the existence of continuous solutions ψ and χ of (11.10). The following lemma gives a sufficient condition on locally compact second countable groups.

Lemma 11.15. Let G be a locally compact second countable group and let $\{\phi_{nj}, 1 \leq j \leq k_n\}$ be a uniformly infinitesimal family of continuous normalised positive definite functions on G, such that $\phi_n = \prod_j \phi_{nj}$ converges pointwise to a continuous positive definite function ϕ. Assume further that there exists a subsequence $(\phi_{n'})$ of (ϕ_n) and a neighbourhood $\mathcal{O}$ of the identity in G such that the function

$$\sup_{n'} |\sum_j (\phi_{n'j} - 1)| \tag{11.11}$$

is bounded on the set $D_o \cap \mathcal{O}$, where D_o is the commutator subgroup of G_o. Then there exists an open subgroup G_o of G, a continuous normalised conditionally positive definite function ψ on G_o, and a continuous homomorphism χ from G_o into the group of complex numbers of modulus 1, such that

$$\phi(g) = \chi(g).\exp \psi(g) \text{ for every } g \varepsilon G_o,$$
$$= 0 \text{ otherwise.}$$

Proof: First of all we note, that by Remark 9.4 and Lemma 11.12 we may always assume that there exists a subsequence $(\phi_{n'})$ of (ϕ_n) such that $\sum_j \{\phi_{n',j}(gh)\phi_{n',j}(g)^{-1}\phi_{n',j}(h)^{-1} - 1\}$ converges pointwise on $G_o \times G_o$ to a continuous kernel $N(g,h)$ satisfying (11.5) for some cocycle δ, and that $\sum_j \{\phi_{n',j} - 1\}$ converges on the commutator subgroup D_o to a function f. We may assume that $(\phi_{n'})$ satisfies (11.11). Then there exists a neighbourhood $\mathcal{O}$ of the identity in G and a constant m_o such that $|f(g)| < m_o$ for every $g \varepsilon D_o \cap \mathcal{O}$. What we have to show is that for every sequence (g_k) in D_o, which converges to e, we have $\lim_k f(g_k) = 0$. Since $f(e) = 0$, this will imply the continuity of f at e, and, together with the continuity of N, this will show that f is continuous on D_o. We use the same argument as in the proof of Theorem 5.4. Assume that $\lim_k f(h_k) = c \neq 0$ for some sequence (h_k) in D_o which converges to e. By Lemma 11.9 we have $\lim_k f(h_k^m) = mc$ for every integer m. Choosing m sufficiently large, we arrive at a contradiction to the assumption that $|f|$ is bounded by m_o in $\mathcal{O} \cap D_o$. So f is continuous. By Lemma 11.11 we can extend f to a continuous function ψ on G_o, which satisfies (11.9). That ψ is conditionally positive definite follows exactly as in the end of the proof of Theorem 11.13. As before we may assume that $\text{Re}\,\psi = \log |\phi|$. The continuity of ψ implies now the continuity of χ . The lemma is proved.

In some cases it is possible to show directly that the functions ψ and χ can be chosen as continuous. We just give three examples.

Theorem 11.16. Let G be a compact second countable group and let $\{\phi_{nj}, 1 \leq j \leq k_n\}$ be a uniformly infinitesimal family of continuous normalised positive definite functions on G. Assume that the sequence $\phi_n = \prod_j \phi_{nj}$ converges pointwise to a continuous positive definite function ϕ on G. Then there exists an open subgroup G_o of G, a continuous normalised conditionally positive definite function ψ on G_o, and a continuous homomorphism χ from G_o into the group of complex numbers of modulus 1, such that

$$\begin{aligned} \phi(g) &= \chi(g) . \exp \psi(g) \text{ for every } g \in G_o, \\ &= 0 \text{ otherwise.} \end{aligned}$$

Proof: This is an immediate consequence of (11.6) and Lemma 11.7.

Theorem 11.17. Let G be a locally compact second countable abelian group, and let $\{\phi_{nj}\}$, ϕ and G_o be defined as in Theorem 11.16. Then there exists a continuous normalised conditionally positive definite function ψ on G_o, and a continuous homomorphism χ of G_o into the group of complex numbers of modulus 1 such that

$$\begin{aligned} \phi(g) &= \chi(g).\exp \psi(g) \text{ for every } g \in G_o \\ &= 0 \text{ otherwise.} \end{aligned}$$

Proof: This follows from Lemma 11.15.

Theorem 11.18. Let G be a locally compact second countable group, and let $\{\phi_{nj}, 1 \leq j \leq k_n\}$ be a uniformly infinitesimal family of continuous normalised positive definite functions on G such that $\phi_n = \prod_j \phi_{nj}$ converges pointwise to a continuous positive definite function ϕ on G. Let $G_o = \{g: \phi(g) \neq 0\}$ and let $\bar{D}_o$ be the closure of the commutator D_o of G_o. Assume the following is fulfilled:

(1) for every compact subset K of $\bar{D}_o$, we have

$$\lim_n \sup_j \sup_{g \in K} |\phi_{nj}(g) - 1| = 0,$$

(2) $\bar{D}_o$ possesses a connected relatively compact neighbourhood $\mathcal{O}$ of the identity (in the subgroup topology).

Then there exists a continuous normalised conditionally positive definite function ψ on G_o and a continuous homomorphism χ from G_o into the group of complex numbers of modulus 1, such that

$$\begin{aligned} \phi(g) &= \chi(g).\exp \psi(g) \text{ for } g \in G_o, \\ &= 0 \text{ otherwise.} \end{aligned}$$

<u>Proof</u>: Choose an integer n_o such that $\sup_j \sup_{g \in \mathcal{O}} |\phi_{nj}(g) - 1| < \frac{1}{2}$ for all $n > n_o$, and define Log $\phi_{nj}(g)$ on the principal branch of the logarithm for all $n > n_o$, j, and $g \in \mathcal{O}$. Put $f_n(g) = \sum_j \text{Log } \phi_{nj}$. Then $\exp f_n(g) = \phi_n(g)$ for all $g \in \mathcal{O}$, $n > n_o$. Since ϕ_n converges pointwise to ϕ, and since ϕ is continuous, ϕ_n converges to ϕ uniformly on every compact subset of G, and hence on $\mathcal{O}$. This is a classical result by Gelfand. The reader can easily verify it using Dixmier, 'Les C^*-algebres', Theorem 13.5.2, and Lebesgue's dominated convergence theorem. From the uniform convergence of the sequence (ϕ_n), the connectedness of $\mathcal{O}$ and the continuity

of f_n it follows that the sequence f_n converges uniformly to a continuous function f on $\mathcal{O}$. But on the other hand we have, by the inequality $|\mathrm{Log}\, x + 1 - x| \leq |1 - x|^2$ for $|1 - x| < \frac{1}{2}$,

$$|f(g) - \sum_j \{\phi_{nj}(g) - 1\}| \leq \lim_n \sup \sum_j |\phi_{nj}(g) - 1|^2 = 0,$$

by (11.7), for every $g \in D_o$. Hence $\lim_n \sup |\sum_j \{\phi_{nj}(g) - 1\}|$ is bounded on $\mathcal{O}$, and we can apply Lemma 11.15 to complete the proof.

§12. Infinitely divisible positive definite functions

Let G be a fixed group.

Definition 12.1. A normalised positive definite function ϕ on G is said to be *infinitely divisible* if, for every integer n, there exists a positive definite function ϕ_n on G such that $\phi = \phi_n^n$. Any positive definite function ϕ_n with $\phi_n^n = \phi$ is called an *n-th root* of ϕ.

Theorem 12.12. Let G be a group and let ϕ be an infinitely divisible positive definite function on G. Then the set $G_o = \{g : \phi(g) \neq 0\}$ is a subgroup of G. Further there exists a unitary representation V of G_o in a Hilbert space H and a map δ from G_o into H such that

$$V_g \delta(h) = \delta(gh) - \delta(g) \quad \text{for all } g, h \in G_o,$$

$$\phi(g_1 g_2) \phi(g_1)^{-1} \phi(g_2)^{-1} = \exp \langle \delta(g_2), \delta(g_1^{-1}) \rangle \quad \text{for all } g_1, g_2 \in G_o.$$

If G is a topological group and if ϕ is a continuous infinitely divisible positive definite function on G which possesses continuous roots, then the representation V of the open subgroup G_o is weakly continuous and the map δ is continuous.

Proof: If $\phi = \phi_n^n$ for every $n = 2,3,\ldots$, we may write $\phi_{nj} = \phi_n$ for $j = 1,\ldots,n$. It is easy to show as before that $G_o = \{g: \phi(g) \neq 0\}$ is a subgroup of G. On G_o, the $\{\phi_{nj}\}$constitute a uniformly infinitesimal family of normalised positive definite functions. An application of Theorem 11.2 and Remark 10.6 completes the proof of this theorem.

In the proof of Theorem 12.2 we have seen that any infinitely divisible positive definite function ϕ on G gives rise to a uniformly infinitesimal family $\{\phi_{nj}\}$ of normalised positive definite functions on the subgroup G_o in a manner described above. We can apply the results of the previous section to get the following:

Theorem 12.3. Let G be a group and letϕ be an infinitely divisible positive definite function on G. Let $G_o = \{g: \phi(g) \neq 0\}$. Then there exists a normalised conditionally positive definite function ψ on G_o and a homomorphism χ from G_o into the multiplicative group of complex numbers of modulus 1, such that the following holds:

$$\begin{aligned} \phi(g) &= \chi(g).\exp \psi(g) \text{ for every } g \in G_o, \\ &= 0 \text{ otherwise.} \end{aligned}$$

If G is in addition locally compact and second countable, if the function ϕ is continuous and has continuous roots ϕ_n, and if there exists a neighbourhood $\mathcal{O}$ of the identity in G and a subsequence $(\phi_{n'})$ of (ϕ_n) such that the function

$$\sup_{n'} n'|\phi_{n'}(g) - 1| \tag{12.1}$$

is bounded on $D_o \cap \mathcal{O}$, where D_o is the commutator subgroup of G_o, then the conditionally positive definite function ψ and the homomorphism χ can both be chosen as continuous.

The following result gives a condition on the topology of G which guarantees the existence of continuous maps ψ and χ. In this context we also refer to Theorem 12.9, where it is shown that under a slightly stronger condition the homomorphism χ can be chosen as $\equiv 1$.

<u>Theorem 12.4</u>. Let G be a locally compact second countable group and let ϕ be a continuous infinitely divisible positive definite function on G which has continuous roots ϕ_n. Let $G_o = \{g: \phi(g) \neq 0\}$ and assume that either G or the closure $\bar{D}_o$ of the commutator D_o of G_o is locally connected. Then there exists a continuous normalised conditionally positive definite function ψ on G_o and a continuous homomorphism χ from G_o into the group of complex numbers of modulus 1 such that

$$\begin{aligned} \phi(g) &= \chi(g).\exp \psi(g) \text{ for every } g\epsilon G_o \\ &= 0 \text{ otherwise.} \end{aligned}$$

Proof: Let $\mathcal{O}$ be a connected neighbourhood of e in G, such that $|\phi(g) - 1| < \frac{1}{2}$ for every $g\in\mathcal{O}$. We define Log $\phi(g)$ on the principal branch of the logarithm for every $g\in\mathcal{O}$. Since both ϕ_n and $\exp \frac{1}{n}$ Log ϕ are continuous n-th roots of ϕ in $\mathcal{O}$, and since they coincide at e, they must coincide in the whole of $\mathcal{O}$ by the connectedness of $\mathcal{O}$. Hence there exists a constant m_o such that

$$\sup_n |\phi_n(g) - 1| \leq \frac{m_o}{n} \cdot |\phi(g) - 1|,$$

for every $g\in\mathcal{O}$. From this we conclude that (12.1) holds in $\mathcal{O}\cap D_o$, where D_o is the commutator in G_o. An application of Theorem 12.3 completes the proof of the case, when G is locally connected. If $\bar{D}_o$ is locally connected, the proof is entirely analogous.

Similarly we have from the Theorems 11.16 and 11.17:

Theorem 12.5. Let G be either a compact second countable group or a locally compact second countable abelian group, and assume ϕ is a continuous normalised infinitely divisible positive definite function on G. Then there exists a continuous normalised conditionally positive definite function ψ on $G_o = \{g: \phi(g)\neq 0\}$ and a continuous homomorphism χ from G_o into the group of complex numbers of modulus 1, such that

$$\begin{aligned} \phi(g) &= \chi(g).\exp \psi(g) \text{ for all } g\in G_o, \\ &= 0 \text{ otherwise.} \end{aligned}$$

Let us finally investigate under which circumstances we can write any continuous normalised infinitely divisible positive definite function on G as the exponential of a continuous normalised conditionally positive definite function. In other words, when can we put $\chi \equiv 1$? Under some assumptions concerning the topology of G we can answer this question.

We shall first prove a few lemmas on the existence of continuous logarithms in connected and locally connected spaces.

Lemma 12.6. Let X be a connected, locally connected, locally compact and second countable topological space. Suppose f is a nowhere vanishing complex valued continuous function such that for some point $x_o \in X$, $f(x_o) = 1$. Let f_n be a sequence of continuous functions on X such that $f_n(x_o) = 1$, $f_n^n(x) = f(x)$ for all $x \in X$ and all n, and $\lim_n f_n(x) = 1$ uniformly on every compact set. Then there exists a continuous function α on X such that $\alpha(x_o) = 0$, $f(x) = \exp \alpha(x)$ and $f_n(x) = \exp \frac{1}{n} \alpha(x)$ for all x and all n.

Proof: We write $F_n(x) = f_{2^n}(x)$, $F_o(x) = f(x)$. Since F_n is continuous and converges uniformly on compact sets to unity, we can choose for each n = 0,1,... a logarithm α_n for F_n such that $\alpha_n(x_o) = 0$, α_n is continuous on a connected neighbourhood C_n of x_o, $\exp \alpha_n(x) = F_n(x)$ for all x, and such that C_n increases to the whole space X. We have

$$F_n^{2^n}(x) = (F_{n+r}^{2^r})^{2^n}(x) = F_o(x) \text{ for all } n,\ r=0,1,2,\ldots, \text{ and } x \in X.$$

Since F_o is continuous and nonvanishing, the ratio $F_{n+r}^{2^r}/F_n$ is a continuous function which takes only a finite number of values.

Since X is connected, $F_{n+r}^{2^r} = F_n$. Thus $\exp 2^r \alpha_{n+r}(x) = \exp \alpha_n(x)$ for all x. Since C_n is connected and α_n and α_{n+r} are continuous on C_n, it follows that $\alpha_n(x) = 2^r \alpha_{n+r}(x)$ for all $x \in C_n$. In other words, $2^n \alpha_n(x) = 2^{n+r} \alpha_{n+r}(x)$ for all $x \in C_n$ and all nonnegative integers n,r. If we put $\alpha(x) = 2^n \alpha_n(x)$ for $x \in C_n$, α is continuous on X and satisfies the requirements of the lemma.

Lemma 12.7. Let X be a topological space satisfying the conditions of Lemma 12.6. Suppose f is a nowhere vanishing continuous complex valued function on X such that $f(x_o) = 1$ for some fixed point $x_o \in X$. Let f_n be a sequence of continuous functions on X such that $f_n(x_o) = 1$ and $f_n^n(x) = f(x)$ for all $x \in X$ and all n. Then f_n converges to unity uniformly on every compact subset of X.

Proof: Let y be any fixed point in X. We choose a connected open neighbourhood N(y) of y such that

$$\operatorname*{Var}_{N(y)} f = \sup_{x_1, x_2 \in N(y)} |f(x_1) - f(x_2)| < \tfrac{1}{2}.$$

If we choose and fix log f(y) somehow, then we can define log f(x) by continuity uniquely in N(y). By the connectedness of N(y) and the continuity of f_n, there exists a constant c(n,y) for every n, such that

$$f_n(x) = \exp \{\tfrac{1}{n} \log f(x) + c(n,y)\} \text{ for } x \in N(y).$$

This implies that $\operatorname*{Var}_{N(y)} f_n$ tends to zero as $n \to \infty$. Thus, for every point $x \in X$, there exists a neighbourhood N(x), such that $\lim_n \operatorname*{Var}_{N(x)} f_n = 0$. If $\lim_n f(x')$ exists for some $x' \in N(x)$, then f_n converges uniformly to the constant $\lim_n f_n(x')$ on N(x). Let now

C(1) be the set of all x for which $\lim_n f_n(x) = 1$. We have just shown that C(1) is open. If $C(1) \neq X$, then there exists at least one y_o in the boundary of C(1), since C(1) is nonempty. We can again find a neighbourhood $N(y_o)$, such that $\operatorname{Var}_{N(y_o)} f_n \to 0$. Since $N(y_o) \cap C(1) \neq \emptyset$, we conlude that $N(y_o) \subset C(1)$. But $N(y_o)$ must have nonempty intersection with the complement of C(1), which yields a contradiction. So we have proved that C(1) = X. That the convergence is uniform on compact subsets is immediate.

__Corollary 12.8.__ Let X be as in Lemma 12.7. Let $x_o \varepsilon X$ be a fixed point and f be a nowhere vanishing complexvalued function on X. Suppose f_n is a sequence of continuous functions on X such that $f_n(x_o) = f(x_o) = 1$ and $f_n^n(x) = f(x)$ for all $x \varepsilon X$ and all n. Then there exists a unique continuous function α such that $\alpha(x_o) = 0$, $f(x) = \exp \alpha(x)$, $f_n(x) = \exp \frac{1}{n} \alpha(x)$ for all $x \varepsilon X$ and all n.

__Theorem 12.9.__ Let G be a connected, locally connected, locally compact and second countable group. Suppose ϕ is a continuous normalised infinitely divisible positive definite function on G. Then there exists a continuous normalised conditionally positive definite function ψ on G such that $\phi(g) = \exp \psi(g)$ for all $g \varepsilon G$. Conversely, if ψ is a conditionally positive definite function which is continuous and normalised, then $\exp \psi(g) = \phi(g)$ for all $g \varepsilon G$ defines an infinitely divisible continuous positive definite function on G.

__Proof__: It is clear from Theorem 12.2 that ϕ does not vanish anywhere. If $\phi = \phi_n^n$ for every n, then by Corollary 12.8, there exists a continuous function ψ on G such that $\psi(e) = 0$ and that $\exp \frac{1}{n} \psi(g) = \phi_n(g)$ for all $g = 2,3,\ldots$, and $\exp \psi(g) = \phi(g)$. It is clear that $n\{\phi_n(g) - 1\}$ is conditionally positive definite and hence is its limit ψ. The converse follows from Lemma 1.7. This completes the proof of the theorem.

Remark 12.10. In classical probability theory, the Levy-Khinchine formula describes all continuous normalised conditionally positive definite functions on the real line. We have so far arrived at a characterisation of such functions in terms of cocycles by combining the results of Theorem 11.2 and 12.2 with the Theorems 11.13, 11.15 - 11.18, and 12.3 - 12.5. In the following sections we will show how the Levy-Khinchine formula arises from our analysis.

Part III. The analysis of cocycles of first order on some special groups

§13. Cocycles of direct integrals of representations

It is clear from the earlier sections that cocycles of first order on separable metric groups with values in a Hilbert space play a very important role in the construction of affine invariant conditionally positive definite kernels and thereby in finding the limits of products of uniformly infinitesimal families of positive definite kernels. Cocycles of first order were also

found to be of fundamental importance for the construction of continuous tensor products of group representations.
The general case when the group G acts on a space X has not been sufficiently analysed. The case when G acts transitively on X can be reduced to the study of the case when G acts on G by left translation. We shall restrict ourselves to this case only in all that follows. We shall furthermore always assume that G is a locally compact second countable group.

Definition 13.1. Let $g \to U_g$ be a weakly continuous (not necessarily unitary) representation of G in a complex separable Hilbert space H. A continuous map $\delta: G \to H$ is called a first order cocycle or simply a cocycle if $U_g \delta(h) = \delta(gh) - \delta(g)$ for all $g, h \in G$. δ is said to be a coboundary if there exists a vector v in H such that $\delta(g) = U_g v - v$ for all $g \in G$.

We shall now reduce the problem of finding the cocycles of an arbitrary unitary representation of G to those of an irreducible representation. We shall make use of some of the standard results in the theory of direct integrals of representations. The reader may refer to G.W.Mackey's Chicago Lecture Notes on group representations.

Theorem 13.2. Let U be a continuous unitary representation of G which is a direct integral $\int U^{\omega} d\mu(\omega)$ over some finite measure space $(\Omega, \mathcal{S}, \mu)$. Suppose δ is a cocycle for U. Then $\delta(g)$ is of the form $\delta(\omega, g)$, where for each fixed ω, $\delta(\omega, \cdot)$ is a continuous cocycle for the representation U^ω, $\omega\in\Omega$.

<u>Proof</u>: Suppose U^ω acts in the Hilbert space H^ω and U acts in $H = \int H^\omega \, d\mu(\omega)$. Then $\delta(g)$ is a function $\delta(\omega,g)$ of two variables $g\in G$, $\omega\in\Omega$, and $\delta(\omega,g)\in H^\omega$. The fact that δ is a continuous cocycle for U implies that

$$U_g\delta(\omega,h) = \delta(\omega,gh) - \delta(\omega,g) \text{ a.e. } \omega(\mu),$$

for every pair $g,h\in G$. We shall assume that almost everywhere statements in Ω and G are with respect to μ and the Haar measure in G respectively. By applying Fubini's theorem to the function

$$F(g,h,\omega) = ||U_g^\omega\delta(\omega,h) - \delta(\omega,gh) + \delta(\omega,g)||^2$$

over every $K\times K\times\Omega$, where K is a compact subset of G, we conclude that

$$U_g^\omega\delta(\omega,h) = \delta(\omega,gh) - \delta(\omega,g) \text{ a.e. } g,h \tag{13.1}$$

for all $\omega\in A$, where A is a set with $\mu(\Omega-A) = 0$. (13.1) implies

$$\delta(\omega,g) = U^\omega_{g^{-1}}\delta(\omega,gh) - U^\omega_{g^{-1}}\delta(\omega,g) \text{ a.e. } g,h \tag{13.2}$$

for all $\omega\in A$. Choose and fix a continuous integrable function $\alpha(g)$ on G which vanishes outside a compact set and for which $\int \alpha(g)\,dg = 1$, where dg stands for integration with respect to the right invariant Haar measure. Then (13.2) implies

$$\delta(\omega,h) = \int\alpha(g)\,\delta(\omega,h)\,dg =$$

$$= \int\alpha(gh^{-1})\,U^\omega_{hg^{-1}}\delta(\omega,g)\,dg - \int\alpha(g)\,U^\omega_{g^{-1}}\delta(\omega,g)\,dg$$

for all $\omega\in A$. For every $\omega\in A$, the right hand side of the above equation is continuous in h. Hence we may assume without loss of generality that $\delta(\omega,h)$ is continuous in h for all $\omega\in\Omega$. In such a case (13.1) holds for all g,h. In other words, $\delta(\omega,.)$ is a continuous cocycle for U^{ω} for every ω. This completes the proof.

§14. Cocycles of induced representations

We shall describe briefly the notion of an induced representation. Let G be a locally compact second countable group acting transitively and continuously on a homogeneous space X. Choose and fix a point $x_o \in X$ and write $K = \{g: gx_o = x_o\}$. K is a closed subgroup of G. It is called the stability subgroup of x_o. Let π be the canonical map $g \to gx_o$ from G onto X. It follows from a theorem of Kuratowski that there exists a one-one borel map ρ from X into G such that $\pi\rho(x) = x$ for all $x \in X$ and $\rho(x_o) = e$. (For a proof see, for example, K.R.Parthasarathy, Probability measures on metric spaces, chapter I.) We choose and fix such a map ρ. For any $g\in G$, $x\in X$, we have $\pi\rho(gx) = \pi(g\rho(x)) = gx$. Hence

$$\rho(gx)^{-1}g\rho(x) \in K \quad \text{for all } x\in X,\ g\in G. \qquad (14.1)$$

We write

$$h(g,x) = \rho(gx)^{-1}g\rho(x) \qquad (14.2)$$

Then

$$h(g_1g_2,x) = h(g_1,g_2x)h(g_2,x) \quad \text{for all } g_1,g_2\in G,\ x\in X. \qquad (14.3)$$

The importance of this functional equation will be seen subsequently. Let λ be any totally finite measure on G equivalent to a Haar measure. Then the measure $\lambda\pi^{-1} = \mu$ has the property of quasi invariance, i.e. $\mu(gE) = 0$ for all g whenever $\mu(E) = 0$. It is a rather elementary theorem that any other quasi invariant measure is equivalent to μ. We shall choose and fix such a quasi invariant measure μ. For any $g\varepsilon G$, let μ^g be the measure defined by the equation $\mu^g(E) = \mu(gE)$ for all borel sets E. Since μ^g and μ are equivalent, the Radon Nikodym derivative

$$\alpha(g,x) = \frac{d\mu}{d\mu^g}(x) \tag{14.4}$$

exists for every g. However, $\alpha(g,x)$ is defined only a.e.$x(\mu)$. It is easy to show that for every $g_1, g_2 \varepsilon G$,

$$\alpha(g_1g_2,x) = \alpha(g_1,g_2x)\alpha(g_2,x) \text{ a.e. } x(\mu). \tag{14.5}$$

Note the similarity between (14.3) and (14.5). Suppose

$$\Sigma(g) = \frac{d_1g}{d_rg}, \qquad \sigma(h) = \frac{d_1h}{d_rh},$$

where d_1g, d_rg, d_1h, d_rh stand for the differentials of the left (right) invariant Haar measures of the groups G and K, respectively, and their quotients denote the Radon Nikodym derivatives. Then $\Sigma$and σ are homomorphisms of the respective groups. It can be shown that there exists a non-negative function β on G such that

$$\beta(gh) = \beta(g)\Sigma(h)\sigma(h)^{-1} \text{ for all } g\varepsilon G,\ h\varepsilon K, \tag{14.6}$$

$$\alpha(g_1,\pi(g_2)) = \beta(g_1g_2)\beta(g_2)^{-1} \text{ a.e.} g_2 \text{ for every } g_1. \tag{14.7}$$

We write

$$\tau(h) = \Sigma(h)^{\frac{1}{2}}\sigma(h)^{-\frac{1}{2}}$$

and call it the Mackey homomorphism of the subgroup K.
Let V be a complex separable Hilbert space with inner product $(.,.)$ and $L_2(\mu,V)$ be the Hilbert space of all maps $f:X\to V$ with the property that they are weakly measurable and that

$$\int (f(x),f(x))\, d\mu(x) < \infty.$$

The inner product in $L_2(\mu,V)$ is defined by

$$\langle f_1,f_2\rangle = \int (f_1(x),f_2(x))\, d\mu(x).$$

If μ is an invariant measure, i.e. if $\mu^g = \mu$ for all $g\varepsilon G$, then the operator U_g: $f(x)\to f(g^{-1}x)$ defines a unitary representation $g\to U_g$ of the group G. Even if μ is not invariant but only quasi invariant, we may modify the definition of U_g by putting U_g: $f(x)\to\alpha(g,g^{-1}x)^{\frac{1}{2}}f(g^{-1}x)$, where α is the function defined by (14.4), and still obtain a unitary representation of G. In defining this representation we have first multiplied f by a scalar function on $G\times X$ which satisfies an equation of the form (14.5) and then moved the point x to $g^{-1}x$ by the element g^{-1} of the group. If instead of a scalar valued function we use a function with values as operators in V and satisfying (14.5), then also we obtain a representation! Let $h\to L_h$ be a unitary representation of the subgroup K in the Hilbert space V. Let

$$C(g,x) = L_{h(g,x)} \tag{14.8}$$

where $h(g,x)$ is defined by (14.2). Then (14.3) implies that

$$C(g_1g_2,x) = C(g_1,g_2x)C(g_2,x) \text{ for all } g_1,g_2\in G,\ x\in X. \qquad (14.9)$$

We define the operator U_g^L in $L_2(\mu,V)$ by

$$(U_g^L f)(x) = \alpha(g,g^{-1}x)^{\frac{1}{2}}\ C(g,g^{-1}x)\ f(g^{-1}x) \qquad (14.10)$$

where α is defined by (14.4) and C by (14.8). The map $g \to U_g^L$ is a unitary representation of G. It is called the induced representation, which is induced by the representation L of the subgroup K. It is a theorem that any general solution of (14.9) is of the form $A(gx)L_{h(g,x)}A(x)^{-1}$ where A is a borel map from X into the group of unitary operators in V and L is a unitary representation of the subgroup K in V. We remark here that if the G action on X is not transitive but ergodic with respect to μ, the problem of solving equation (14.9) is still open. Let now U^L be the induced representation defined by (14.10) where C is given by (14.8). We write

$$D(g,x) = \alpha(g,x)^{\frac{1}{2}}\ C(g,x), \qquad (14.11)$$

$$\tilde{D}(g_1,g_2) = D(g_1,\pi(g_2)). \qquad (14.12)$$

If we write

$$\theta(g) = \beta(g)^{\frac{1}{2}}\ L_{(\rho\pi(g\ )^{-1}g\rho\pi(g)} \qquad (14.13)$$

where β is determined by (14.6) and (14.7) and ρ and π are the maps described earlier, then

$$\tilde{D}(g_1,g_2) = \theta(g_1g_2)\theta(g_2)^{-1} \text{ for all } g_1g_2\in G, \qquad (14.14)$$

$$\theta(gh) = \theta(g)\tau(h)\, L_h \text{ for all } g\varepsilon G,\ h\varepsilon K, \tag{14.15}$$

where τ is the Mackey homomorphism of K. We write

$$\hat{L}_h = \tau(h)\, L_h \tag{14.16}$$

and observe that the map $h \to \hat{L}_h$ is also a representation of K. Now suppose that δ is a cocycle for the representation U^L defined by (14.10). $\delta(g)$ is of the form $\delta(g,x)$, where $x \to \delta(g,x)$ is a borel map from X into V, satisfying the equation

$$D(g_1,g_1^{-1}x)\delta(g,g_1^{-1}x) = \delta(g_1g,x) - \delta(g_1,x) \text{ a.e.} x(\mu) \tag{14.17}$$

for all g_1,g in G. For any borel function f on X we define $\tilde{f}$ on G by $\tilde{f}(g) = f(\pi(g))$. Almost everywhere statements on G and X will be with respect to the Haar measure and μ respectively. Then (14.17) implies

$$\tilde{D}(g_1,g_1^{-1}g_2)\tilde{\delta}(g,g_1^{-1}g_2) = \tilde{\delta}(g_1g,g_2) - \tilde{\delta}(g_1,g_2) \text{ a.e.} g_2,$$

for every g,g_1. Putting $g_1^{-1}g_2 = g'$ and using Fubini's theorem we find that, for some fixed $g_o\varepsilon G$,

$$\tilde{\delta}(g,g') = \theta(g')\theta(g_o)^{-1}\{\tilde{\delta}(g_og'^{-1}g,g_o) - \tilde{\delta}(g_og'^{-1},g_o)\} \quad \text{a.e.} g,g'. \tag{14.18}$$

We write

$$\psi(g') = \theta(g_o)^{-1}\tilde{\delta}(g_og'^{-1},g_o).$$

Then

$$\tilde{\delta}(g,g') = \theta(g')\{\psi(g^{-1}g') - \psi(g')\} \text{ a.e.} g,g'. \tag{14.19}$$

Since $\tilde{\delta}(g,g'h) = \tilde{\delta}(g,g')$ a.e.g,g' and $\theta(g'h) = \theta(g')\hat{L}_h$, for every $h\varepsilon K$, we have

$$\psi(gh^{-1}) = \hat{L}_h\psi(g) + \varepsilon(h) \text{ a.e.g} \tag{14.20}$$

for every $h\varepsilon K$. This implies that the map ε from K into V is a cocycle for the representation $\hat{L}$ of K. (14.18) and (14.20) yield after some straightforward computation,

$$\tilde{\delta}(g,g') = \theta(g\rho(g^{-1}\pi(g')))\psi(\rho(g^{-1}\pi(g'))) -$$

$$- \theta(\rho\pi(g'))\psi(\rho\pi(g')) + \theta(\rho\pi(g'))\varepsilon((\rho\pi(g')^{-1}g\rho(g^{-1}\pi(g')))$$
$$\text{a.e.g,g'}.$$

If we now write $f(x) = \theta(\rho(x))\psi(\rho(x))$ and observe that $\theta(\rho(x)) = \beta(\rho(x))^{\frac{1}{2}}$, we obtain

$$\delta(g,x) = \{\alpha(g,g^{-1}x)^{\frac{1}{2}}C(g,g^{-1}x)f(g^{-1}x) - f(x)\}$$
$$+ \beta(\rho(x))^{\frac{1}{2}}\varepsilon(\rho(x)^{-1}g\rho(g^{-1}x)) \text{ a.e.x,a.e.g.} \tag{14.21}$$

We recall that β is determined from (14.6) and (14.7). If we denote the right hand side of (14.21) by $\delta'(g,x)$ and extend the operation U_g^L of (14.10) to all borel functions, we see that both δ and δ' satisfy the cocycle identity. This implies that the set of points g such that $\delta(g,x) = \delta'(g,x)$ a.e.x is a group. Hence the two agree everywhere. Thus we have proved the following theorem:

<u>Theorem 14.1</u>. Let G be a locally compact second countable group acting transitively on a homogeneous space X. For some

fixed point $x_0 \in X$, let K be the stability subgroup of x_0. Suppose L is a unitary representation of K in a separable Hilbert space V and τ is the Mackey homomorphism of K. Let U^L be the induced representation of G defined by (14.10). If δ is a cocycle for the representation U^L, then $\delta(g)(x) = \delta(g,x)$ is given by the equation

$$\delta(g,x) = \{\alpha(g,g^{-1}x)^{\frac{1}{2}}C(g,g^{-1}x)f(g^{-1}x) - f(x)\}$$

$$+ \beta(\rho(x))^{\frac{1}{2}}\varepsilon(\rho(x)^{-1}g\rho(g^{-1}x)) \text{ a.e.} x \qquad (14.22)$$

for every g, where f is a borel map from X into V, ε is a cocycle for the representation $\hat{L}$ of K, defined by $\hat{L}_h = \tau(h)L_h$, and α, β, ρ and C are the quantities entering the equations (14.1) - (14.8). Conversely, if f is a borel map from X into V and ε is a cocycle for the representation L such that the right hand side of (14.22) is a continuous map from G into $L_2(\mu,V)$, then $\delta(g)$ defined by the equation $\delta(g)(x) = \delta(g,x)$ in (14.22) is a continuous cocycle for U^L.

Remark 14.2. If ε is a coboundary for L, then $\delta(g,x)$ is of the form $\alpha(g,g^{-1}x)^{\frac{1}{2}}C(g,g^{-1}x)f(g^{-1}x) - f(x)$ for some borel map f from X into V.

§15. Cocycles for compact groups

Theorem 15.1. Let G be a compact second countable group and U be a unitary representation of G in a separable Hilbert space H. Then every cocycle of U is a coboundary.

Proof: Suppose δ is a cocycle for U. Then

$$U_g\delta(h) = \delta(gh) - \delta(g) \text{ for all } g,h\in G.$$

Integrating the above equation with respect to the normalised Haar measure and writing $v = \int -\delta(h)\, dh$, we obtain $\delta(g) = U_g v - v$ for all g. This completes the proof.

Remark 15.2. For any coboundary $\delta(g) = U_g v - v$, we have

$$\langle\delta(g_2),\delta(g_1^{-1})\rangle = \{\langle U_{g_1g_2}v,v\rangle - \langle v,v\rangle\} - \{\langle U_{g_1}v,v\rangle - \langle v,v\rangle\}$$
$$- \{\langle U_{g_2}v,v\rangle - \langle v,v\rangle\}.$$

This shows that the function ψ appearing in Theorem 11.15 and in Theorem 12.5 (for compact G) is of the form $\psi(g) = \langle U_g v,v\rangle - \langle v,v\rangle$ for some continuous unitary representation U of the open subgroup G_o and some vector x in the space H, where U acts.

Remark 15.3. A modification of the proof of Theorem 15.1 shows that any bounded cocycle on an amenable group is a coboundary. In particular this holds for abelian groups and compact groups.

§16. Cocycles of abelian groups

We shall now discuss the situation when G is a locally compact second countable abelian group. We shall denote the group operation in G by +. Let Γ denote the character group of G with the identity 1. For any $g\in G$, $\gamma\in\Gamma$, let (g,γ) denote the value of the character γ at the point g. An irreducible unitary representation of G is determined by an element $\gamma \varepsilon \Gamma$. This representation is simply multiplication by (g,γ) on the complex plane. A cocycle δ for the representation of G determined by γ is a continuous complex valued function $\delta(g)$ on G satisfying

$$(g,\gamma)\delta(h) = \delta(g+h) - \delta(g) \text{ for all } g,h\varepsilon G. \tag{16.1}$$

Theorem 16.1. Let G be a locally compact second countable abelian group and γ a character of G which is not identically equal to 1. Then every cocycle for the one dimensional represen-tagion of G corresponding to γ is a coboundary. If $\gamma \equiv 1$, then every cocycle for γ is a continuous homomorphism from G into the complex plane.

Proof: The second part of the theorem is obvious from (16.1). To prove the first part we have to solve equation (16.1). Let G_o be the subgroup $\{g: (g,\gamma) = 1\}$. Then (16.1) implies

$$\delta(g_1+g_2) = \delta(g_1) + \delta(g_2) \text{ for all } g_1\varepsilon G_o. \tag{16.2}$$

If we choose $g_2 \not\varepsilon G_o$, (16.1) implies

$$\delta(g_1+g_2) = (g_2,\gamma)\delta(g_1) + \delta(g_2).$$

The above two equations and the fact that $(g_2,\gamma) \neq 1$ imply that $\delta(g_1) = 0$ for all $g_1 \varepsilon G_o$. If we choose $g,h \not\varepsilon G_o$, we obtain from (16.1)

$$(g,\gamma)\delta(h) + \delta(g) = \delta(g+h) = (h,\gamma)\delta(g) + \delta(h).$$

Hence $\delta(g)/\{(g,\gamma) - 1\}$ is a constant c for all $g \notin G_0$. In other words $\delta(g) = c\{(g,\gamma) - 1\}$ for all g, i.e. δ is a coboundary. This completes the proof.

We shall now obtain a formula for the function $N(g_1,g_2) = \langle\delta(g_2),\delta(g_1^{-1})\rangle$, where δ is a first order cocycle for an arbitrary representaion U of G in a separable Hilbert space H with inner product $\langle .,.\rangle$. By (11.6), N is a second order cocycle on G×G. Suppose U is the trivial representation of G in H, i.e. $U_g = I$ for all g, where I is the identity operator in H. Then δ is simply an additive homomorphism from G into H. In this case we can write

$$\begin{aligned} N(g_1,g_2) = &- \tfrac{1}{2} \{||\delta(g_1+g_2)||^2 - ||\delta(g_1)||^2 - ||\delta(g_2)||^2\} \\ &+ 2 \tfrac{1}{2i} \{\langle\delta(g_2),\delta(g_1)\rangle - \langle\delta(g_1),\delta(g_2)\rangle\}. \end{aligned} \tag{16.3}$$

We observe that $\frac{1}{2i} \{\langle\delta(g_2),\delta(g_1)\rangle - \langle\delta(g_1),\delta(g_2)\rangle\}$ is a real skew symmetric biadditive form on G×G.

Before proceeding to discuss the case when U is a nontrivial representation we shall state a lemma, for a proof of which we refer to K.R.Parthasarathy, 'Probability measures on metric spaces', Chapter IV, Lemma 5.3.

Lemma 16.2. There exists a function $m: G\times\Gamma\to R$, where R denotes the real line, such that the following properties hold:

(1) m(.,.) is continuous in both variables,

(2) $\sup_{g\in K}\sup_{\gamma\in\Gamma}|m(g,\gamma)| < \infty$ for every compact set $K\subset G$,

(3) $m(g_1+g_2,\gamma) = m(g_1,\gamma) + m(g_2,\gamma)$ for all $g_1,g_2\in G$, $\gamma\in\Gamma$,

(4) for any compact set $K\subset G$ there exists a neighbourhood N_K of the identity in Γ such that $(g,\gamma) = \exp i\, m(g,\gamma)$ for all $g\in K$, $\gamma\in N_K$,

(5) for any compact set $K\subset G$, $\lim_{\gamma\to 1}\sup_{g\in K}|m(g,\gamma)| = 0$.

We remark that properties (2), (4) and (5) imply that for any compact set $K\subset G$, there exists a neighbourhood N_K of the identity in Γ such that

$$|(g,\gamma) - 1 - i\, m(g,\gamma)| \leq c_1\, m(g,\gamma)^2, \tag{16.4}$$

$$|(g,\gamma) - 1 - i\, m(g,\gamma)| \leq c_2\, \{1 - \mathrm{Re}\,(g,\gamma)\}, \tag{16.5}$$

for all $g\in K$, $\gamma\in N_K$, where c_1 and c_2 are constants depending on K.

Now suppose that U is a nontrivial representation of G in H. It is a well known result that U can be expressed as a direct integral of one dimensional representations. We assume that the trivial character does not enter this direct integral decomposition. Then there exists a measure space $(\Omega,\mathcal{S},\mu)$ and a measurable map $\tau:\Omega\to G$ such that $\tau(\omega)$ is a nontrivial character for every ω and that $U = \int \tau(\omega)\, d\mu(\omega)$, where $\tau(\omega)$ is considered as the one dimensional representation acting in the complex plane. Suppose δ is a cocycle for U. By Theorem 13.2, $\delta(g)$ is of the form $\delta(\omega,g)$, where $\delta(\omega,g)$ is a cocycle for $\tau(\omega)$ for every ω. Hence, since $\delta(\omega,g) = c(\omega)\{(g,\tau(\omega))-1\}$ for all ω by Theorem 16.1,

$$N(g_1,g_2) = \int |c(\omega)|^2 \{(g_2,\tau(\omega)) - 1\}\{(g_1,\tau(\omega)) - 1\} d\mu(\omega)$$
$$\text{for all } g_1,g_2. \tag{16.6}$$

If we write $\mu_1(E) = \int_E |c(\omega)|^2 d\mu(\omega)$ for $E\in\mathcal{S}$ and $F = \mu_1\tau^{-1}$, then (16.6) can be written as

$$N(g_1,g_2) = \int \{(g_2,\gamma) - 1\}\{(g_1,\gamma) - 1\} dF(\gamma). \tag{16.7}$$

We have

$$||\delta(g)||^2 = \int |(g,\gamma) - 1|^2 dF(\gamma)$$
$$= 2 \int \{1 - \text{Re } (g,\gamma)\} dF(\gamma).$$

Since δ is a continuous map from G into H, it follows that the right hand side should be finite for every g and that F must have finite mass outside every neighbourhood of the identity in Γ. If we write

$$t(g,\gamma) = (g,\gamma) - 1 - i\, m(g,\gamma),$$

it follows from the properties of the measure F, property (2) of m in Lemma 16.2 and (16.5) that $\int t(g,\gamma) dF(\gamma)$ exists and is a continuous function on G. From property (3) of m in Lemma 16.2 and (16.7) we have

$$N(g_1,g_2) = \int \{t(g_1+g_2,\gamma) - t(g_1,\gamma) - t(g_2,\gamma)\} dF(\gamma).$$

Let r be a continuous homomorphism from G into the real line. If we write

$$\psi(g) = \int t(g,\gamma) dF(\gamma) + i\, r(g) \tag{16.8}$$

then

$$N(g_1,g_2) = \psi(g_1+g_2) - \psi(g_1) - \psi(g_2), \tag{16.9}$$

$$\psi(g^{-1}) = \overline{\psi(g)}. \tag{16.10}$$

The positive definiteness of the kernel $<\delta(g_1),\delta(g_2)> = N(g_2^{-1},g_1)$ on G×G and Lemma 1.7 implies that ψ is a conditionally positive definite function on G. It is immediate that ψ is normalised. Conversely if ψ is a normalised continuous conditionally positive definite function satisfying (16.9), where $N(g_1,g_2) = <\delta(g_2),\delta(g_1^{-1})>$ for some cocycle δ for a representaion which does not contain the trivial representation, then ψ is of the form (16.8).

Since any unitary representation of G can be written as the direct sum of a trivial representation and of a representation which does not contain the trivial representation we have the following theorem:

Theorem 16.3. Let U be a unitary representation of a locally compact second countable abelian group G in a separable Hilbert space H, and let δ be a continuous first order cocycle for U. Then there exists a continuous homomorphism θ from G into H, a continuous homomorphism r from G into the real line, and a σ-finite measure F on the character group Γ of G such that

$$<\delta(g_2),\delta(g_1^{-1})> = \{\psi(g_1+g_2) - \psi(g_1) - \psi(g_2)\}$$
$$+ \tfrac{1}{2} \{<\delta(g_2), \delta(g_1)> - <\delta(g_1),\delta(g_2)>\}$$
$$\text{for all } g_1,g_2 \in G,$$

where

$$\psi(g) = \int \{(g,\gamma) - 1 - i\, m(g,\gamma)\}\, dF(\gamma) + i\, r(g)$$

$$- \tfrac{1}{2}\, ||\theta(g)||^2 \quad \text{for all } g\epsilon G.$$

The function m is defined on $G \times \Gamma$ by Lemma 16.2. Further the measure F has finite mass outside every neighbourhood of the identity and $\int \{1 - \mathrm{Re}\,(g,\gamma)\}\, dF(\gamma) < \infty$ for every $g\epsilon G$.

As a consequence of the Theorems 11.17 and 12.3 we obtain

Remark 16.4. On a locally compact second countable abelian group G, let $\{\phi_{nj}\}$ be a uniformly infinitesimal family of continuous normalised positive definite functions. Suppose $\phi_n = \prod_j \phi_{nj}$ converges pointwise to a continuous positive definite function ϕ. Then there exists an open subgroup G_o of G, a continuous character χ on G_o, and a continuous normalised conditionally positive definite function ψ on G_o, which is of the form described in Theorem 16.2, such that $\phi = \chi.\exp \psi$ on G_o and zero outside G_o. For any infinitely divisible continuous normalised positive definite function on G the same result holds good by Theorem 10.5 and Theorem 16.3.

Theorem 16.3 together with Remark 16.4 describes the limiting laws of the socalled central limit theorems of probability theory. For further information the reader may refer to 'Probability measures on metric spaces' by the first author.

§17. Cocylces of nilpotent Lie groups

Lemma 17.1. Let G be a connected nilpotent Lie group and let χ be a nontrivial one dimensional unitary representation of G. Then any cocycle of χ is a coboundary.

Proof: Let δ be a cocycle for χ. Then $\chi(g_1)\delta(g_2) = \delta(g_1g_2) - \delta(g_1)$. Consider the map τ: $g \to \begin{pmatrix}\chi(g) & \delta(g)\\ 0 & 1\end{pmatrix}$. Then τ is a homomorphism from G into the solvable group of all matrices of the form $\begin{pmatrix}e^{i\theta} & z\\ 0 & 1\end{pmatrix}$, where z lies in the complex plane and θ in the interval $[0,2\pi]$. The image $\tau(G)$ is a nilpotent subgroup. Since χ is a nontrivial character and G is connected, $\chi(g)$ takes all the values in the unit circle of the complex plane. Hence $\tau(G)$ contains a one parameter subgroup of the form $\begin{pmatrix}e^{it} & z_t\\ 0 & 1\end{pmatrix}$, where t runs over the real line. z_t satisfies the equation $e^{it}z_s + z_t = z_{t+s}$. In other words z_t is a cocycle for the representation $t \to e^{it}$ of the real line. Hence by Theorem 16.1 $z_t = c(e^{it} - 1)$ for all t, where c is a constant. The nilpotency of $\tau(G)$ implies that $\tau(G)$ is just this one parameter subgroup. Hence $\delta(g) = c.\{\chi(g) - 1\}$. This completes the proof of the lemma.

Lemma 17.2. Let G be a connected nilpotent Lie group and let Z_o be its centre. Suppose K is a connected and closed subgroup of G containing Z_o. Let χ be a one dimensional unitary representation of K, which is nontrivial on Z_o. Then every cocycle of the induced representation U^{χ} is a coboundary.

Proof: The homogeneous space X = G/K admits an invariant measure and hence its Mackey homomorphism is trivial. By Lemma 17.1

any cocycle of χ is a coboundary. Hence by Theorem 14.1 and Remark 14.2, any cocycle for U^{χ} is given by the equation

$$\delta(g,x) = \chi(\rho(x)^{-1}g\cdot\rho(g^{-1}x))f(g^{-1}x) - f(x) \text{ for all } g, \qquad (17.1)$$

where f is a borel function on G/K and ρ is as in Theorem 14.1. Hence

$$\delta(g,x) = \chi(g)f(x) - f(x) \text{ for all } g\varepsilon Z_o.$$

Since $\delta(g,.)$ is square integrable on X and $\chi(g) \neq 1$ for some $g\varepsilon Z_o$, it follows that f is square integrable. Hence (17.1) becomes

$$\delta(g,x) = U_g^{\chi}f\ (x) - f(x) \text{ for all } g.$$

In other words δ is a coboundary. This completes the proof of the lemma.

Lemma 17.3. Let G be a connected nilpotent Lie group with centre Z_o. Suppose U is a nontrivial irreducible unitary representation of G, which is trivial on Z_o. Then any cocycle δ of U is invariant under translations by elements of Z_o, i.e. $\delta(gh) = \delta(g)$ for all $g\varepsilon G$, $h\varepsilon Z_o$.

Proof: The cocycle equation for δ implies that

$$U_g\delta(h) + \delta(g) = \delta(gh) = \delta(hg) = U_h\delta(g) + \delta(h)$$
$$\text{for all } g\varepsilon G,\ h\varepsilon Z_o.$$

Since U_h is the identity operator, we have $U_g\delta(h) = \delta(h)$ for all g. Since U is irreducible and nontrivial, $\delta(h) = 0$ for all $h\varepsilon Z_o$. Thus $\delta(gh) = \delta(g)$ for all $g\varepsilon G$, $h\varepsilon Z_o$. This completes the proof.

It is a theorem of Kirillov that every irreducible unitary representation of a connected nilpotent Lie group is induced by a one dimensional representation of a closed and connected subgroup containing the centre. If this one dimensional representation is trivial on the centre, then the induced representation is trivial on the centre and hence by Lemma 17.3, the problem of finding the cocycles of G can be reduced to finding those of G/Z_0. This induction procedure together with Lemma 17.2 yields the following theorem:

Theorem 17.4. Let G be a connected nilpotent Lie group and let U be a nontrivial irreducible unitary representation of G. Then every cocycle of U is a coboundary.

§18. Cocyles of semi-simple Lie groups

Let G be a connected semi-simple Lie group and let K be its maximal compact subgroup. Let G = KAN be its Iwasawa decomposition and let M be the maximal torus in K. We shall find the cocycles of those representations which are induced by a one dimensional representation of the subgroup MAN. By theorem 14.1, the cocycles of such representations are determined by cocycles of one dimensional representations of MAN.

Lemma 18.1. Let L be a locally compact second countable group with closed subgroups $B, N \subset L$, where $B \cap N = \{e\}$, B is abelian, N is normal and BN = L. Let χ be a one dimensional representation of L such that $\chi(bn) = \chi(b)$ for all $b\epsilon B$, $n\epsilon N$. Then any cocycle δ of χ is given by

$$\delta(bn) = \chi(b)\eta(n) + c.\{\chi(b) - 1\} \text{ for all } b\epsilon B,\ n\epsilon N,$$

where c is a constant and η is a continuous additive homomorphism of N into the complex plane which satisfies the condition

$$\eta(bnb^{-1}) = \chi(b)\eta(n) \text{ for all } b\epsilon B,\ n\epsilon N.$$

Proof: This follows from a straightforward analysis of the cocycle identity.

Remark 18.2. We shall now apply Lemma 18.1 to the case when $L = AN$, where A is the Cartan subgroup of a simply connected complex semi-simple Lie group G and N is the nilpotent subgroup occuring in the Iwasawa decomposition of G. Let $\mathfrak{g}$, $\mathfrak{a}$, $\mathfrak{n}$ be the Lie algebras of G,A,N respectively. Let P_+ be a system of positive roots of $\mathfrak{g}$, considered as a complex Lie algebra. Let $\{X_\alpha,\ \alpha\epsilon P_+\}$ be the positive root vectors. We assume that $\mathfrak{n}$ is the vector space spanned by the X_α. Let χ be a one dimensional representation of A and let η be a continuous complex valued homomorphism of N such that $\chi(a)\eta(n) = \eta(ana^{-1})$ for all $a\epsilon A, n\epsilon N$. Suppose $d\eta(X_{\alpha_o}) \neq 0$ for some α_o. Since $d\eta$ is a Lie algebra homomorphism, α_o must be a simple root. Further for all complex numbers c, we have

$$\chi(\exp D)\eta(\exp c\, X_{\alpha_o}) = \eta(\exp c\{\exp \alpha_o(D)\}X_{\alpha_o}),\ D\epsilon\,\mathfrak{a}.$$

Hence

$$\chi(\exp D) = \exp \alpha_o(D) \text{ for all } D\epsilon\,\mathfrak{a}.$$

If χ is not given by the above equation, η is identically 0. If $\chi(\exp D) \equiv \exp \alpha_o(D)$ for some simple positive root, then $\eta(\exp \sum_{\alpha\varepsilon P_+} c_\alpha X_\alpha) = c'.c_{\alpha_o}$ for all complex numbers c_α, $\alpha\varepsilon P_+$, where c' is some constant.

Using this, one can find all cocycles of all irreducible unitary representations of, for example, SL(2,C).

References

[1] Araki, H. and Woods, E.J.: Complete Boolean algebras of type I factors, Publications of RIMS, Kyoto University, Ser.A, Vol.2, No.2, 157-242 (1966).

[2] Araki, H.: Factorisable representations of current algebra, Publications of RIMS, Kyoto University, Ser.A, Vol.5, No.3, 361-422 (1970).

[3] Bargmann, V.: Unitary ray representations of continuous groups, Ann.Math. Vol.59, 1-46 (1954).

[4] Cook, J.M.: The mathematics of second quantization, Trans. Amer.Math.Soc. 74, 222-245 (1953).

[5] Dixmier, J.: Les C^*-algebres et leurs representations, Gauthier-Villars, Paris 1969.

[6] Gangolli, R.: Positive definite kernels on homogeneous spaces, Ann.Inst.H.Poincare B, Vol.3, 121-225 (1967).

[7] Gnedenko, B.V. and Kolmogorov, A.N.. Limit distributions for sums of independent variables, Cambridge 1954.

[8] Johansen, S.: An application of extreme point methods to the representation of infinitely divisible distributions, Z.Wahrscheinlichkeitstheorie verw.Geb. 5, 304-316 (1966).

[9] Kleppner, A.: Multipliers on abelian groups, Math.Ann. 158, 11-34 (1965).

[10] Mackey, G.W.: Theory of group representations (lecture notes), Dept.Math., University of Chicago (1955).

[11] Moore, C.C.: Extensions and low dimensional cohomology theory of locally compact groups I, Trans.Amer.Math. Soc. 113, 40-63 (1964).

[12] Parthasarathy, K.R.: Probability measures on metric spaces, Academic Press, New York 1967.

[13] ——— : Multipliers on locally compact groups, Lecture Notes in Mathematics, Springer, 1969.

[14] ——— : Infinitely divisible representations and positive definite functions on a compact group, Comm.Math.Phys. Vol.16, 148-156 (1970).

[15] ——— and Schmidt, K.: Infinitely divisible projective representations, cocycles and Levy-Khinchine-Araki formula on locally compact groups, Research Report 17, Manchester-Sheffield School of Probability and Statistics, (1970).

[16] ——— and Schmidt, K.: Factorisable representations of current groups and the Araki-Woods imbedding theorem, Acta Mathematica, Vol.128, 53-71 (1972).

[17] Schmidt, K.: Limits of uniformly infinitesimal families of projective representations of locally compact groups, Math.Ann. 192, 107-118 (1971).

[18] Segal, I.E.: Mathematical problems of relativistic physics, Amer.Math.Soc. 1963.

[19] Streater, R.F.: Current commutation relations, continuous tensor products, and infinitely divisible group representations, Rendiconti di Sc.Int.di Fisica E.Fermi, Vol XI, 247-263 (1969).

Vol. 146: A. B. Altman and S. Kleiman, Introduction to Grothendieck Duality Theory. II, 192 pages 1970. DM 18,–

Vol. 147: D. E. Dobbs, Cech Cohomological Dimensions for Commutative Rings. VI, 176 pages 1970 DM 16,–

Vol. 148: R Azencott, Espaces de Poisson des Groupes Localement Compacts IX, 141 pages. 1970. DM 16,–

Vol. 149: R. G. Swan and E. G. Evans, K-Theory of Finite Groups and Orders. IV, 237 pages 1970. DM 20,–

Vol. 150: Heyer, Dualität lokalkompakter Gruppen. XIII, 372 Seiten 1970 DM 20,–

Vol 151: M Demazure et A Grothendieck, Schémas en Groupes I. (SGA 3). XV, 562 pages. 1970 DM 24,–

Vol. 152: M Demazure et A. Grothendieck, Schémas en Groupes II. (SGA 3). IX, 654 pages. 1970 DM 24,–

Vol. 153: M Demazure et A Grothendieck, Schémas en Groupes III. (SGA 3). VIII, 529 pages 1970 DM 24,–

Vol 154: A. Lascoux et M. Berger, Variétés Kähleriennes Compactes VII, 83 pages 1970 DM 16,–

Vol 155: Several Complex Variables I, Maryland 1970. Edited by J. Horváth. IV, 214 pages 1970 DM 18,–

Vol. 156: R. Hartshorne, Ample Subvarieties of Algebraic Varieties XIV, 256 pages. 1970 DM 20,–

Vol. 157: T. tom Dieck, K H Kamps und D. Puppe, Homotopietheorie. VI, 265 Seiten. 1970. DM 20,–

Vol. 158: T G Ostrom, Finite Translation Planes IV 112 pages. 1970. DM 16,–

Vol 159: R Ansorge und R. Hass Konvergenz von Differenzenverfahren für lineare und nichtlineare Anfangswertaufgaben. VIII, 145 Seiten 1970 DM 16,–

Vol. 160: L. Sucheston, Constributions to Ergodic Theory and Probability. VII, 277 pages. 1970. DM 20,–

Vol. 161: J. Stasheff, H-Spaces from a Homotopy Point of View. VI, 95 pages. 1970. DM 16,–

Vol 162: Harish-Chandra and van Dijk, Harmonic Analysis on Reductive p-adic Groups IV, 125 pages 1970 DM 16,–

Vol 163: P. Deligne, Equations Différentielles à Points Singuliers Reguliers III, 133 pages. 1970. DM 16,–

Vol. 164: J. P Ferrier, Seminaire sur les Algebres Complètes II, 69 pages. 1970. DM 16,–

Vol. 165: J M Cohen, Stable Homotopy. V, 194 pages 1970 DM 16,–

Vol. 166: A. J Silberger, PGL_2 over the p-adics: its Representations, Spherical Functions, and Fourier Analysis VII, 202 pages. 1970. DM 18,–

Vol. 167: Lavrentiev, Romanov and Vasiliev, Multidimensional Inverse Problems for Differential Equations. V, 59 pages. 1970 DM 16,–

Vol. 168: F P Peterson, The Steenrod Algebra and its Applications: A conference to Celebrate N. E Steenrod's Sixtieth Birthday VII, 317 pages 1970. DM 22,–

Vol 169: M Raynaud, Anneaux Locaux Henséliens V, 129 pages 1970 DM 16,–

Vol 170: Lectures in Modern Analysis and Applications III. Edited by C. T Taam VI, 213 pages. 1970. DM 18,–

Vol. 171: Set-Valued Mappings, Selections and Topological Properties of 2^X. Edited by W. M. Fleischman. X, 110 pages 1970. DM 16,–

Vol. 172: Y.-T. Siu and G. Trautmann, Gap-Sheaves and Extension of Coherent Analytic Subsheaves. V, 172 pages 1971. DM 16,–

Vol. 173: J. N. Mordeson and B. Vinograde, Structure of Arbitrary Purely Inseparable Extension Fields IV, 138 pages. 1970. DM 16,–

Vol. 174: B. Iversen, Linear Determinants with Applications to the Picard Scheme of a Family of Algebraic Curves VI, 69 pages 1970. DM 16,–

Vol 175: M Brelot, On Topologies and Boundaries in Potential Theory. VI, 176 pages. 1971 DM 18,–

Vol. 176: H. Popp, Fundamentalgruppen algebraischer Mannigfaltigkeiten. IV, 154 Seiten. 1970. DM 16,–

Vol 177: J Lambek, Torsion Theories, Additive Semantics and Rings of Quotients. VI, 94 pages. 1971. DM 16,–

Vol. 178: Th. Bröcker und T. tom Dieck, Kobordismentheorie. XVI, 191 Seiten. 1970. DM 18,–

Vol. 179: Seminaire Bourbaki – vol. 1968/69. Exposés 347-363. IV. 295 pages 1971 DM 22,–

Vol. 180: Séminaire Bourbaki – vol. 1969/70 Exposés 364-381. IV, 310 pages. 1971. DM 22,–

Vol 181: F. DeMeyer and E. Ingraham, Separable Algebras over Commutative Rings. V, 157 pages. 1971. DM 16,–

Vol. 182: L D. Baumert. Cyclic Difference Sets VI, 166 pages. 1971. DM 16,–

Vol. 183: Analytic Theory of Differential Equations. Edited by P. F. Hsieh and A W J Stoddart VI, 225 pages 1971 DM 20,–

Vol 184: Symposium on Several Complex Variables, Park City, Utah, 1970 Edited by R. M. Brooks. V, 234 pages. 1971. DM 20,–

Vol. 185: Several Complex Variables II, Maryland 1970 Edited by J. Horváth. III, 287 pages. 1971. DM 24,–

Vol. 186: Recent Trends in Graph Theory Edited by M. Capobianco/ J. B. Frechen/M Krolik VI, 219 pages 1971 DM 18.–

Vol. 187: H. S. Shapiro, Topics in Approximation Theory. VIII, 275 pages. 1971. DM 22,–

Vol. 188: Symposium on Semantics of Algorithmic Languages. Edited by E. Engeler VI, 372 pages. 1971. DM 26,–

Vol. 189: A. Weil, Dirichlet Series and Automorphic Forms. V, 164 pages. 1971. DM 16,–

Vol. 190: Martingales. A Report on a Meeting at Oberwolfach, May 17-23, 1970. Edited by H. Dinges. V, 75 pages. 1971. DM 16,–

Vol. 191: Séminaire de Probabilités V Edited by P A. Meyer. IV, 372 pages. 1971. DM 26,–

Vol. 192: Proceedings of Liverpool Singularities – Symposium I. Edited by C. T. C. Wall V, 319 pages. 1971. DM 24,–

Vol. 193: Symposium on the Theory of Numerical Analysis Edited by J. Ll. Morris. VI, 152 pages. 1971. DM 16,–

Vol. 194: M. Berger, P. Gauduchon et E. Mazet. Le Spectre d'une Variété Riemannienne VII, 251 pages. 1971. DM 22,–

Vol. 195: Reports of the Midwest Category Seminar V. Edited by J W Gray and S. Mac Lane. III, 255 pages. 1971. DM 22,–

Vol. 196: H-spaces – Neuchâtel (Suisse)- Août 1970. Edited by F. Sigrist, V, 156 pages. 1971. DM 16,–

Vol 197: Manifolds – Amsterdam 1970. Edited by N. H Kuiper V, 231 pages 1971 DM 20,–

Vol 198: M Hervé, Analytic and Plurisubharmonic Functions in Finite and Infinite Dimensional Spaces VI, 90 pages. 1971. DM 16.–

Vol. 199: Ch J. Mozzochi, On the Pointwise Convergence of Fourier Series VII, 87 pages. 1971 DM 16,–

Vol 200: U Neri, Singular Integrals VII, 272 pages. 1971. DM 22,–

Vol. 201: J. H. van Lint, Coding Theory. VII, 136 pages 1971 DM 16,–

Vol 202: J. Benedetto, Harmonic Analysis on Totally Disconnected Sets. VIII, 261 pages 1971. DM 22,–

Vol. 203: D. Knutson, Algebraic Spaces. VI, 261 pages. 1971. DM 22,–

Vol 204: A Zygmund, Intégrales Singulières. IV, 53 pages. 1971 DM 16,–

Vol. 205: Séminaire Pierre Lelong (Analyse) Année 1970. VI, 243 pages. 1971. DM 20,–

Vol. 206: Symposium on Differential Equations and Dynamical Systems. Edited by D. Chillingworth. XI, 173 pages. 1971. DM 16,–

Vol 207: L. Bernstein, The Jacobi-Perron Algorithm – Its Theory and Application IV, 161 pages 1971 DM 16,–

Vol. 208: A. Grothendieck and J. P. Murre, The Tame Fundamental Group of a Formal Neighbourhood of a Divisor with Normal Crossings on a Scheme VIII, 133 pages. 1971. DM 16,–

Vol. 209: Proceedings of Liverpool Singularities Symposium II. Edited by C. T. C. Wall. V, 280 pages. 1971. DM 22,–

Vol. 210: M. Eichler, Projective Varieties and Modular Forms III, 118 pages. 1971. DM 16,–

Vol. 211: Théorie des Matroïdes. Edité par C. P. Bruter. III, 108 pages. 1971. DM 16,–

Vol. 212: B. Scarpellini, Proof Theory and Intuitionistic Systems. VII, 291 pages. 1971. DM 24,–

Vol. 213: H Hogbe-Nlend, Théorie des Bornologies et Applications. V, 168 pages. 1971. DM 18,–

Vol. 214: M. Smorodinsky, Ergodic Theory, Entropy. V, 64 pages. 1971. DM 16,–

Lecture Notes in Mathematics

Comprehensive leaflet on request

Vol. 74: A. Fröhlich, Formal Groups. IV, 140 pages. 1968. DM 16,–

Vol. 75: G. Lumer, Algèbres de fonctions et espaces de Hardy. VI, 80 pages. 1968. DM 16,–

Vol. 76: R. G. Swan, Algebraic K-Theory. IV, 262 pages. 1968. DM 18,–

Vol. 77: P.-A. Meyer, Processus de Markov: la frontière de Martin. IV, 123 pages. 1968. DM 16,–

Vol. 78: H. Herrlich, Topologische Reflexionen und Coreflexionen. XVI, 166 Seiten. 1968. DM 16,–

Vol. 79: A. Grothendieck, Catégories Cofibrées Additives et Complexe Cotangent Relatif. IV, 167 pages. 1968. DM 16,–

Vol. 80: Seminar on Triples and Categorical Homology Theory. Edited by B. Eckmann. IV, 398 pages. 1969. DM 20,–

Vol. 81: J.-P. Eckmann et M. Guenin, Méthodes Algébriques en Mécanique Statistique. VI, 131 pages. 1969. DM 16,–

Vol. 82: J. Wloka, Grundräume und verallgemeinerte Funktionen. VIII, 131 Seiten. 1969. DM 16,–

Vol. 83: O. Zariski, An Introduction to the Theory of Algebraic Surfaces. Second Printing. IV, 100 pages. 1972. DM 16,–

Vol. 84: H. Lüneburg, Transitive Erweiterungen endlicher Permutationsgruppen. IV, 119 Seiten. 1969. DM 16,–

Vol. 85: P. Cartier et D. Foata, Problèmes combinatoires de commutation et réarrangements. IV, 88 pages. 1969. DM 16,–

Vol. 86: Category Theory, Homology Theory and their Applications I. Edited by P. Hilton. VI, 216 pages. 1969. DM 16,–

Vol. 87: M. Tierney, Categorical Constructions in Stable Homotopy Theory. IV, 65 pages. 1969. DM 16,–

Vol. 88: Séminaire de Probabilités III. IV, 229 pages. 1969. DM 18,–

Vol. 89: Probability and Information Theory. Edited by M. Behara, K. Krickeberg and J. Wolfowitz. IV, 256 pages. 1969. DM 18,–

Vol. 90: N. P. Bhatia and O. Hajek, Local Semi-Dynamical Systems. II, 157 pages. 1969. DM 16,–

Vol. 91: N. N. Janenko, Die Zwischenschrittmethode zur Lösung mehrdimensionaler Probleme der mathematischen Physik. VIII, 194 Seiten. 1969. DM 16,80

Vol. 92: Category Theory, Homology Theory and their Applications II. Edited by P. Hilton. V, 308 pages. 1969. DM 20,–

Vol. 93: K. R. Parthasarathy, Multipliers on Locally Compact Groups. III, 54 pages. 1969. DM 16,–

Vol. 94: M. Machover and J. Hirschfeld, Lectures on Non-Standard Analysis. VI, 79 pages. 1969. DM 16,–

Vol. 95: A. S. Troelstra, Principles of Intuitionism. II, 111 pages. 1969. DM 16,–

Vol. 96: H.-B. Brinkmann und D. Puppe, Abelsche und exakte Kategorien, Korrespondenzen. V, 141 Seiten. 1969. DM 16,–

Vol. 97: S. O. Chase and M. E. Sweedler, Hopf Algebras and Galois theory. II, 133 pages. 1969. DM 16,–

Vol. 98: M. Heins, Hardy Classes on Riemann Surfaces. III, 106 pages. 1969. DM 16,–

Vol. 99: Category Theory, Homology Theory and their Applications III. Edited by P. Hilton. IV, 489 pages. 1969. DM 24,–

Vol. 100: M. Artin and B. Mazur, Etale Homotopy. II, 196 Seiten. 1969. DM 16,–

Vol. 101: G. P. Szegö et G. Treccani, Semigruppi di Trasformazioni Multivoche. VI, 177 pages. 1969. DM 16,–

Vol. 102: F. Stummel, Rand- und Eigenwertaufgaben in Sobolewschen Räumen. VIII, 386 Seiten. 1969. DM 20,–

Vol. 103: Lectures in Modern Analysis and Applications I. Edited by C. T. Taam. VII, 162 pages. 1969. DM 16,–

Vol. 104: G. H. Pimbley, Jr., Eigenfunction Branches of Nonlinear Operators and their Bifurcations. II, 128 pages. 1969. DM 16,–

Vol. 105: R. Larsen, The Multiplier Problem. VII, 284 pages. 1969. DM 18,–

Vol. 106: Reports of the Midwest Category Seminar III. Edited by S. Mac Lane. III, 247 pages. 1969. DM 16,–

Vol. 107: A. Peyerimhoff, Lectures on Summability. III, 111 pages. 1969. DM 16,–

Vol. 108: Algebraic K-Theory and its Geometric Applications. Edited by R. M. F. Moss and C. B. Thomas. IV, 86 pages. 1969. DM 16,–

Vol. 109: Conference on the Numerical Solution of Differential Equations. Edited by J. Ll. Morris. VI, 275 pages. 1969. DM 18,–

Vol. 110: The Many Facets of Graph Theory. Edited by G. Chartrand and S. F. Kapoor. VIII, 290 pages. 1969. DM 18,–

Vol. 111: K. H. Mayer, Relationen zwischen charakteristischen Zahlen. III, 99 Seiten. 1969. DM 16,–

Vol. 112: Colloquium on Methods of Optimization. Edited by N. N. Moiseev. IV, 293 pages. 1970. DM 18,–

Vol. 113: R. Wille, Kongruenzklassengeometrien. III, 99 Seiten. 1970. DM 16,–

Vol. 114: H. Jacquet and R. P. Langlands, Automorphic Forms on GL (2). VII, 548 pages. 1970. DM 24,–

Vol. 115: K. H. Roggenkamp and V. Huber-Dyson, Lattices over Orders I. XIX, 290 pages. 1970. DM 18,–

Vol. 116: Séminaire Pierre Lelong (Analyse) Année 1969. IV, 195 pages. 1970. DM 16,–

Vol. 117: Y. Meyer, Nombres de Pisot, Nombres de Salem et Analyse Harmonique. 63 pages. 1970. DM 16,–

Vol. 118: Proceedings of the 15th Scandinavian Congress, Oslo 1968. Edited by K. E. Aubert and W. Ljunggren. IV, 162 pages. 1970. DM 16,–

Vol. 119: M. Raynaud, Faisceaux amples sur les schémas en groupes et les espaces homogènes. III, 219 pages. 1970. DM 16,–

Vol. 120: D. Siefkes, Büchi's Monadic Second Order Successor Arithmetic. XII, 130 Seiten. 1970. DM 16,–

Vol. 121: H. S. Bear, Lectures on Gleason Parts. III, 47 pages. 1970. DM 16,–

Vol. 122: H. Zieschang, E. Vogt und H.-D. Coldewey, Flächen und ebene diskontinuierliche Gruppen. VIII, 203 Seiten. 1970. DM 16,–

Vol. 123: A. V. Jategaonkar, Left Principal Ideal Rings. VI, 145 pages. 1970. DM 16,–

Vol. 124: Séminare de Probabilités IV. Edited by P. A. Meyer. IV, 282 pages. 1970. DM 20,–

Vol. 125: Symposium on Automatic Demonstration. V, 310 pages. 1970. DM 20,–

Vol. 126: P. Schapira, Théorie des Hyperfonctions. XI, 157 pages. 1970. DM 16,–

Vol. 127: I. Stewart, Lie Algebras. IV, 97 pages. 1970. DM 16,–

Vol. 128: M. Takesaki, Tomita's Theory of Modular Hilbert Algebras and its Applications. II, 123 pages. 1970. DM 16,–

Vol. 129: K. H. Hofmann, The Duality of Compact Semigroups and C*-Bigebras. XII, 142 pages. 1970. DM 16,–

Vol. 130: F. Lorenz, Quadratische Formen über Körpern. II, 77 Seiten. 1970. DM 16,–

Vol. 131: A. Borel et al., Seminar on Algebraic Groups and Related Finite Groups. VII, 321 pages. 1970. DM 22,–

Vol. 132: Symposium on Optimization. III, 348 pages. 1970. DM 22,–

Vol. 133: F. Topsøe, Topology and Measure. XIV, 79 pages. 1970. DM 16,–

Vol. 134: L. Smith, Lectures on the Eilenberg-Moore Spectral Sequence. VII, 142 pages. 1970. DM 16,–

Vol. 135: W. Stoll, Value Distribution of Holomorphic Maps into Compact Complex Manifolds. II, 267 pages. 1970. DM 18,–

Vol. 136: M. Karoubi et al., Séminaire Heidelberg-Saarbrücken-Strasbourg sur la K-Théorie. IV, 264 pages. 1970. DM 18,–

Vol. 137: Reports of the Midwest Category Seminar IV. Edited by S. MacLane. III, 139 pages. 1970. DM 16,–

Vol. 138: D. Foata et M. Schützenberger, Théorie Géométrique des Polynômes Eulériens. V, 94 pages. 1970. DM 16,–

Vol. 139: A. Badrikian, Séminaire sur les Fonctions Aléatoires Linéaires et les Mesures Cylindriques. VII, 221 pages. 1970. DM 18,–

Vol. 140: Lectures in Modern Analysis and Applications II. Edited by C. T. Taam. VI, 119 pages. 1970. DM 16,–

Vol. 141: G. Jameson, Ordered Linear Spaces. XV, 194 pages. 1970. DM 16,–

Vol. 142: K. W. Roggenkamp, Lattices over Orders II. V, 388 pages. 1970. DM 22,–

Vol. 143: K. W. Gruenberg, Cohomological Topics in Group Theory. XIV, 275 pages. 1970. DM 20,–

Vol. 144: Seminar on Differential Equations and Dynamical Systems, II. Edited by J. A. Yorke. VIII, 268 pages. 1970. DM 20,–

Vol. 145: E. J. Dubuc, Kan Extensions in Enriched Category Theory. XVI, 173 pages. 1970. DM 16,–

Please turn over